GODWIN STUDY GUIDES

# MATERIALS
# AND STRUCTURES

*Godwin Study Guides*

GODWIN STUDY GUIDES

# MATERIALS AND STRUCTURES

## M. J. SMITH
M.Sc., C.Eng., M.I.C.E., M.I.Struct.E.
*Senior Lecturer at the Polytechnic of the South Bank*

Second edition

## GEORGE GODWIN LIMITED

The book publishing subsidiary
of The Builder Group

First published 1970 by Macdonald & Evans Ltd
in the series *Examination Subjects for Engineers and Builders*
Reprinted 1974, 1977
Second edition published 1980 by George Godwin Ltd
as a *Godwin Study Guide*

© George Godwin 1980

George Godwin Limited
The book publishing subsidiary
of The Builder Group
1–3 Pemberton Row
Red Lion Court
Fleet Street
London EC4

**British Library Cataloguing in Publication Data**

Smith, Michael John
  Materials and structures. – 2nd ed. – (Godwin Study Guides).
  1. Structures, Theory of
  I. Title
  624′.17          TA645
  ISBN 0 7114 5639 9

Filmset by Northumberland Press Ltd, Gateshead, Tyne and Wear
Printed in Great Britain by Fletcher & Son Ltd, Norwich

# GENERAL INTRODUCTION

THIS series was originally designed as an aid to students studying for technical examinations, the aim of each book being to provide a clear concise guide to the *basic principles* of the subject, reinforced by worked examples carefully selected to illustrate the text. The success of the series with students has justified the original aim, but it became apparent that qualified professional engineers in mid-career were also finding the books useful.

In recognition of this need, the books in the series have been enlarged to cover a wider range of topics, whilst maintaining the concise form of presentation.

It is our belief that this increase in content should help students to see their study material in a more practical context without detracting from the value of the book as an aid to passing examinations. Equally, it is believed that the additional material will present a more complete picture to professional engineers of topics which they have not had occasion to use since completing their original studies.

A list of other books in the series is given at the front of this book. Further details may be obtained from the publishers.

M. J. Smith
*General Editor*

# AUTHOR'S PREFACE TO THE SECOND EDITION

THIS book has been prepared as an elementary text in strength of materials and theory of structures for civil and structural engineers. This standard also covers the requirements in this subject for architects, quantity surveyors and students in other branches of engineering. The questions have been taken from a wide range of examination papers set over many years, but the standard is consistent with the Technician Education Council's Higher Diploma and Certificate (H.T.D. and H.T.C.) together with B.Sc. and C.E.I. examinations, all at first year level.

Although the book has been written primarily to assist students to pass an examination in "theory," an attempt has also been made to give a basic idea of the approach to structural design, which for most architects, surveyors and builders should be sufficient in itself. The civil and structural engineering student will, of course, proceed to more advanced work, but a thorough grasp of the principles laid down here will be essential if he wishes fully to understand advanced structural design. This book may be taken as an introduction to the companion volumes *Theory of Structures* and *Advanced Theory of Structures* in this series.

The design of structures in practice may be taken in two parts. First, it is required to determine the forces acting on a structure and the distribution of those forces within the material of which it is composed. This may be called the "theory of structures." Second, from a knowledge of this material and its ability to withstand the forces applied to it, an estimate of the size of structural members required may be made. This is broadly the scope of "strength of materials."

The external forces acting on any structural member constitute the *load*, which is normally made up of stored material, machinery, people and the weight of the building itself. Some of this load may be permanent or dead load, and some intermittent or live load, but in the early stages of study it is easier to consider all load as dead load. Live load acts in a similar way to dead load, but it is the task of the designer to determine its distribution. These external forces may be applied in such a way as to cause the structural member to compress, extend, buckle, bend, shear, twist or behave in a combination of these ways. In this book each type of behaviour is considered in turn.

To keep the book a reasonable size, only typical structural problems have been dealt with, and the rare types of question on springs and hoop stresses have been omitted. This new edition, however, has

vii

been enlarged to include simple analysis of reinforced concrete beams, three-pin arches and an introduction to influence lines. It has also been revised to bring it into line with current practice.

The author would like to thank the following examining bodies for permission to use questions from their examination papers:

The Institution of Structural Engineers
*Theory of structures and strength of materials, graduateship.*
Council of Engineering Institutions
*Strength of materials and theory of structures,* part 2.
The Institution of Civil Engineers
*Theory of structures,* part 2.
Inner London Education Authority
*Strength of materials* and *theory of structures,* Higher National Certificate.
University of London
*Properties of materials and stress analysis,* B.Sc. (Eng.) part 1.
*Theory of structures,* B.Sc. (Eng.) part 2 (Civil).
City and Guilds of London Institute.
*Structural engineering,* intermediate and final.
The Institute of Quantity Surveyors
*Theory of structures,* second examination.
Royal Institute of British Architects
*Building science (structure),* intermediate.

In all cases, where applicable, the questions used have been converted to SI units. The author accepts full responsibility for the conversion and also for the worked solutions.

This book should be found adequate for all the above examinations as well as the early years of B.Sc., H.N.D. and T.E.C. in civil engineering.

September 1979                                                   M.J.S.

# CONTENTS

CHAPTER 1

# ELASTICITY

## LOAD

LOAD may be defined as any *external* force acting on a material. In any structure, all loads acting on it must balance (*see Equilibrium on page 21*).

## STRESS

Any material will deform under load, and when this deformation takes place *internal* forces in the material resist it. These internal forces are called stresses. The force transmitted across any section divided by the area of that section is the *intensity* of stress, but for brevity it is normally referred to simply as stress. In this chapter only tensile stress, compressive stress and shear stress will be considered.

### Tensile stress
Fig. 1 is a simple example of tensile stress. A bar of uniform cross-sectional area $A$ is fixed at its upper end, and carries an axial load $W$ at its lower end. To support this load an internal force $F$ is required, which is equal in magnitude and opposite in direction to $W$. The intensity of tensile stress $f_t$ is $F/A$. However, it is far easier to measure $W/A$, which is numerically equal to $F/A$.

### Compressive stress
A simple example of compressive stress is shown in Fig. 2. A column of uniform cross-sectional area $A$ carries an axial load $W$. To support this load an internal force $F$ is required. The intensity of compressive stress $f_c$ is $F/A$, which again is easier to measure as $W/A$.

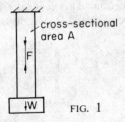

FIG. 1

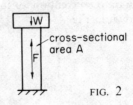

FIG. 2

1

It should be noted that in both these examples the bar and the column have been assumed to be weightless. In practice, if load $W$ is large this assumption will cause little error. However, the stress can be determined at any cross-section, and the weight of column above the section (or weight of rod below it) should be added to load $W$ to to give the exact stress at the section considered.

In tensile and compressive stress the force is at right angles to the cross-section under consideration, and these are known as direct stresses. Further examples of them will be found in Chapter 2, in connection with simple frameworks.

**Shear stress**

A shear stress acts parallel to the section under consideration. An example of shear stress is given in Fig. 3. Two plates fixed together by a bolt of cross-sectional area $A$ support a load $W$. The upper plate is fixed to a rigid support. The figure shows a cross-section through the system. The plates will be subject to direct tensile stress and will have to transfer the load $W$ from one plate to the other. To do this, an internal force $Q$ is required along section $Y–Y$ of the bolt.

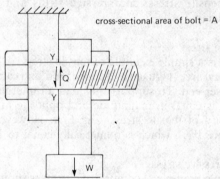

cross-sectional area of bolt = A

FIG. 3

Intensity of shear stress across $Y–Y = f_s = Q/A$, which again may be equated to $W/A$. Note that in this case the internal force is parallel to the section under consideration.

In all the cases mentioned, the internal force has been indicated by a single arrow. In fact for each section to be in balance there must be an internal force in the opposite direction. This leads to the standard convention for indicating the type of stress on diagrams as shown in Fig. 4.

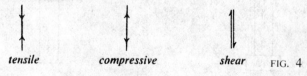

*tensile*        *compressive*        *shear*        FIG. 4

Care should be taken with units. The units of stress are load/unit area, i.e. $N/mm^2$ (newtons per square millimetre).

SPECIMEN QUESTION 1

A square mass concrete column with 600 mm sides and 2500 mm high carries an axial "load" of 600 000 kg. If the concrete weighs 2200 $kg/m^3$, what is the stress at the base of the column?

SOLUTION

$$\text{Applied load} = 600\,000 \times 9\cdot 8$$
$$= 5\,880\,000$$
$$\text{own weight of concrete} = 0\cdot 6 \times 0\cdot 6 \times 2\cdot 5 \times 2200 \times 9\cdot 8$$
$$= \underline{\phantom{00}19\,400}$$
$$\text{Total load:} = 5\,899\,400 \text{ newtons}$$
$$\text{Stress at base of concrete} = \frac{5\,899\,400}{600 \times 600}$$
$$= 16\cdot 4 \text{ N/mm}^2$$

SPECIMEN QUESTION 2

A mild steel bar 25 mm in diameter has a permissible stress of 138 $N/mm^2$ in tension. If the bar is to be used to resist tensile force, what is the load it can carry?

SOLUTION

In this case the bar is assumed stressed to its maximum permissible tensile stress $p_t = (\text{load/area})$.

$$\text{Load} = p_t \times \text{area of bar}$$
$$= 138 \times \frac{\pi \times (25)^2}{4}$$
$$= 67\,700 \text{ newtons (6900 kg)}$$

SPECIMEN QUESTION 3

Two steel plates are lapped together with three black bolts. If the permissible shear stress in the bolts is 80 $N/mm^2$ and the plates are to carry a 9000 kg "load" in tension, what diameter bolts are to be used?

SOLUTION

$$\text{Area of bolt steel required} = \frac{\text{load}}{p_q}$$
$$= \frac{9000 \times 9\cdot 8}{80}$$
$$= 1102\cdot 5 \text{ mm}^2$$

or $1102\cdot 5/3 = 367\cdot 5$ $mm^2$ for each bolt. Therefore $(\pi d^2)/4 = 367\cdot 5$, or $d = 21\cdot 6$ mm. In practice three 22 mm bolts would be used.

## Friction grip bolts

Although black bolts or machined bolts are still used on occasion in structural steelwork construction, friction grip bolts are now in more common use. These friction grip bolts are not designed on the basis of the shear strength of the bolt, but on the friction between the lapping steel plates resisting shearing force. This friction equals $\mu F_p$ where $\mu$ is the coefficient of friction between steel plates (0·45) and $F_p$ is the proof load applied by the bolt at right angles to the plates. Hence friction grip bolts are designed upon the basis of their tensile strength.

# STRAIN

When a material is subject to stress it will deform. The measure of this deformation is known as strain. In tension and compression the strain may be defined as the variation in length per unit length.

Fig. 5(a) shows a bar of length $L$ unloaded. When a load $W$ is applied (*see* Fig. 5(b)) the bar will stretch by a small amount $\delta L$. Tensile strain $e_t = \delta L/L$. In compression the bar or column would be compressed by an amount $\delta L$, and compressive strain $e_t$ would be $\delta L/L$ again.

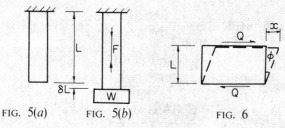

FIG. 5(a)      FIG. 5(b)      FIG. 6

Fig. 6 shows a material subject to shear stress which will deform as indicated. The shear strain $\phi$ is defined as the angular displacement produced by shear stress, i.e. shear strain $= x/L$, since for small angles $\tan \phi \simeq \phi$ the shear strain $= x/L = \phi$.

It should be noted that strain is a ratio of lengths, and therefore will have no units.

SPECIMEN QUESTION 4

If the column in specimen question 1 was found to compress by 0·425 mm, what would be the strain in the concrete?

SOLUTION

Compressive strain $e_t = 0.425/2500 = \underline{\underline{0.17 \times 10^{-3}}}$

# ELASTICITY

Any material will deform under load, and a measurement of strain can be made. If, after removing the load, the material returns to its original shape it is said to be *elastic*. A limiting value of load will be found at which the strain does not completely disappear with the removal of the load. The value of the stress corresponding to this load is called the *elastic limit*.

## Tensile testing (mild steel)

The elastic limit and many other points may readily be seen by carrying out a tensile test. There are many types of apparatus on the market for measuring elasticity accurately, but for simplicity only a simple test will be described. It should be noted that it is difficult to obtain accurate results from this experiment.

A thin mild steel wire is suspended from a rigid support and has a collar for attaching weights at its lower end. The wire is marked at two points and an accurate gauge is fixed beside it to measure the distance between these points. A gauge is required to measure the diameter of the wire.

After measuring the distance between the marked points (initial length *L*) and the diameter of the wire, a small load is placed gently on the collar. The distance between the marked points and the diameter of the wire are measured again. This procedure can then be repeated, gradually increasing the load until the wire stretches for an appreciable amount and finally breaks. (A rubber mat under the wire—and care to keep toes out of reach—are sensible precautions here.)

In the initial stages the load should be removed after each test and it will be observed that the wire returns to its original length. Eventually, on removing the load, a small permanent set will remain. If great care is taken and the load increment is small enough, shortly after this stage it may be observed that the wire continues to extend even on removal of some of the load. However in such a simple experiment it is extremely difficult to locate this point.

The results of this test should be recorded and the stress and strain in the wire calculated for each load increment. A graph of stress against strain is then plotted and a curve, as shown in Fig. 7, is obtained. A number of important basic assumptions are taken from this graph, as follows:

## Hooke's law

Hooke stated that if a material is loaded without exceeding its elastic limit, then the deformation produced is proportional to the load producing it, i.e. deformation $\propto$ load.

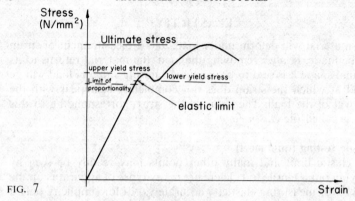

FIG. 7

But deformation $\propto$ strain and load $\propto$ stress, therefore stress $\propto$ strain,

$$\text{or} \qquad \frac{\text{stress}}{\text{strain}} = \text{constant } (E)$$

From the graph for mild steel it can be seen that this law applies, since a straight line is obtained up to the limit of proportionality (not the elastic limit). The constant $E$ is known as *Young's modulus of elasticity*, and is of considerable importance in the following chapters. The units of this modulus are the same as for stress, i.e. load/unit area.

### Limit of proportionality

This is the stress at which Hooke's law ceases to apply. The material may still be in an elastic state beyond this limit.

### Elastic limit

This is the stress at which a permanent set remains in the material after removal of the load. After this point the material retains some elasticity but its behaviour becomes increasingly plastic.

### Yield point

This is the stress at which there is a marked increase in strain. A slight drop in load at this point should cause further strain, giving a lower yield point. With the simple experiment described, this lower point is difficult to locate.

After yielding the material reaches the plastic stage and further loading will cause the specimen to "neck" and finally fracture.

### Maximum stress (ultimate stress)

This is taken as the ratio of the load when necking commences to the *original* cross-sectional area.

*Permissible working stress and factor of safety*

In practical conditions the exact loading on a structure cannot be determined accurately and therefore the material of the structure should not be designed to withstand its maximum stress. To overcome this a factor of safety is applied such that:

$$\text{permissible working stress} = \frac{\text{maximum stress}}{\text{factor of safety}}$$

A more practical limitation for elastic design would be the yield stress, and then:

$$\text{permissible working stress} = \frac{\text{yield stress}}{\text{factor of safety}}$$

This factor of safety depends largely on the material under consideration.

SPECIMEN QUESTION 5

The following figures give observations from a tensile test on a round piece of mild steel 25 mm diameter and 195 mm between gauge points.

| "Load" (Mg): | 5 | 10 | 15 | 16 | 16·5 | 17 |
|---|---|---|---|---|---|---|
| Extension (mm): | 0·009 | 0·019 | 0·029 | 0·034 | 0·046 | 0·078 |

| "Load" (Mg): | 18 | 19 | 20 | 21 | 22 | 23 |
|---|---|---|---|---|---|---|
| Extension (mm): | 0·084 | 0·091 | 0·098 | 0·107 | 0·124 | 0·149 |

| "Load" (Mg): | 24 | 25 | 25·5 | 23 |
|---|---|---|---|---|
| Extension (mm): | 0·188 | 0·239 | 0·395 | 0·493 |

Assuming the change in cross-sectional area to be negligible in the early part of the test, plot the load–extension curve and determine the modulus of elasticity, yield stress, maximum stress and percentage elongation. With a factor of safety of 2 applied to the yield stress, what is the permissible working stress?

SOLUTION

Fig. 8 is the plotted curve from these results. The question asks for a load–extension curve, which, as cross-sectional area normally varies only slightly, serves the same purpose as a stress–strain curve. The graph has been plotted in SI units. Note that Mg loads have been converted to kN (by multiplying by 9·8 m/s$^2$).

Cross-sectional area of bar $= (\pi \times 25^2)/4 = 491$ mm$^2$

Modulus of elasticity $= $ stress/strain
(up to the elastic limit)
$= (140/491) \div (0·025/195)$
$= \underline{2200 \text{ kN/mm}^2}$

Yield stress $= 161\,000/491 = \underline{330 \text{ N/mm}^2}$

Maximum stress $= 250\,000/491 = \underline{510 \text{ N/mm}^2}$

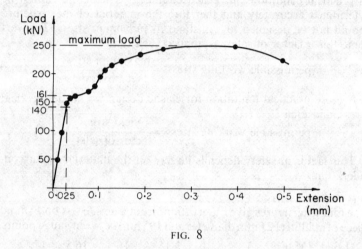

FIG. 8

The percentage elongation is defined as the total increase in gauge length at fracture expressed as a percentage of the original length. The percentage contraction is the reduction in cross-sectional area at the neck (measured after fracture) expressed as a percentage of the original area. These two figures give a measure of the *ductility* of the material.

In specimen question 5:

the percentage elongation = $(0.494/195) \times 100 = \underline{\underline{0.25\%}}$

permissible working stress = $330/2 = \underline{\underline{165 \text{ N/mm}^2}}$

**Proof stress (non-ferrous metals)**
Many non-ferrous metals do not give a clearly defined elastic limit, or yield point. Fig. 9 shows a typical stress–strain curve for a non-ferrous material (curve 0A). If the specimen is loaded beyond the elastic limit (to say, point B on the graph) and then the load gradually removed, a curve OB–BC is obtained where the slope of AB is almost parallel to the early part of the curve OD. The offset OC will be the permanent set in the material.

In practice the material is loaded to failure and the curve OA plotted (*see* Fig. 10). Then a percentage permanent set (0.1% for aluminium alloys) is measured along the strain axis to point C, and a line drawn through C parallel to the early part of the stress–strain curve. Where this line cuts the stress–strain curve at E gives a

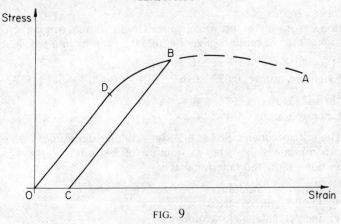

FIG. 9

measure of the proof stress (0·1% proof stress). This is the stress used by the designer for determining permissible working stress, i.e.:

permissible working stress = proof stress/factor of safety.

The elastic limit may be taken in a similar manner, using line FG through 0·02% permanent set to give the 0·02% elastic limit. For the yield stress, use line HJ, giving the 0·2% yield stress.

*Note:* 0·1% permanent set = (0·1/100) × original length
∴ equivalent strain = 0·001 $L/L$ = 0·001

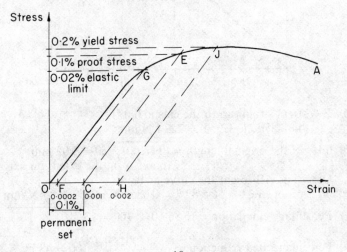

FIG. 10

SPECIMEN QUESTION 6

A non-ferrous metal test piece, gauge length 50 mm, original cross-sectional area 82·5 mm$^2$, gave the following results in a tensile test:

| "Load" (Mg): | 2 | 3 | 4 | 4·5 | 5 | 5·5 |
|---|---|---|---|---|---|---|
| Extension (mm): | 0·053 | 0·080 | 0·107 | 0·120 | 0·140 | 0·172 |

| "Load" (Mg): | 5·75 | 6 |
|---|---|---|
| Extension (mm): | 0·195 | 0·23 |

The test specimen failed at 6·5 Mg, with an extension of 6·86 mm and a minimum diameter at fracture of 7·11 mm. Plot a load–extension graph and determine:

(a) the elastic modulus;
(b) the 0·1% proof stress;
(c) the percentage elongation;
(d) the percentage area reduction.

SOLUTION

The load–extension graph is shown in Fig. 11 (Mg converted to kN).

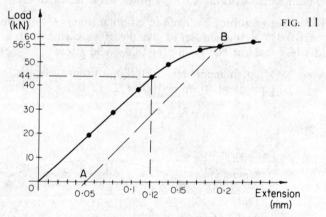

FIG. 11

(a) $E$ = stress/strain (up to the elastic limit)
     = (44/82·5) ÷ (0·12/50) = <u>222 kN/mm$^2$</u>

(b) 0·1% of the original length = (0·1/100) × 50 = 0·05 mm
                                     (point A on the extension axis).
     Draw line AB to cut the curve at B.
     0·1% proof stress = 56·5/82·5 = 0·685 kN/mm$^2$ = <u>685 N/mm$^2$</u>

(c) Percentage elongation = (6·86/50) × 100 = <u>13·7%</u>

(d) Percentage area reduction = $\dfrac{\pi \times (7\cdot11^2/4)}{82\cdot5} \times 100$ = <u>48·1%</u>

In modern design of reinforced concrete the steel reinforcement is assumed stressed beyond its elastic limit, to the 0·2% proof stress and a load factor of 1·15 is used. This gives a more economic use of material.

The permissible working stress of most materials can be found in some manner similar to those described for metals. For concrete, a material commonly used by structural engineers, a compressive test is used, since concrete has only negligible tensile resistance. However, the important relationship to remember and fully understand is:

$$E = \frac{\text{stress}}{\text{strain}} = \frac{\text{load}}{\text{area}} \div \frac{\text{change in length}}{\text{original length}}$$

SPECIMEN QUESTION 7

In a compression test a concrete cube of side 150 mm is loaded with 3·5 Mg. If $E$ for concrete is 140 kN/mm² what would be the percentage compression of the cube at this load?

$$E = \frac{\text{stress}}{\text{strain}} = \frac{\text{load}}{\text{area}} \div \frac{\text{change in length}}{\text{original length}}$$

$$140 = \frac{3 \cdot 5 \times 10^3 \times 9 \cdot 8}{150 \times 150} \times \frac{150}{\delta L}$$

$$\delta L = \frac{3 \cdot 5 \times 10^3 \times 9 \cdot 8}{150 \times 140} = 1 \cdot 6 \text{ mm}$$

$$\therefore \text{ percentage compression} = \frac{1 \cdot 6}{150} \times 100 = \underline{1 \cdot 07\%}$$

## COMPOUND BARS

Any tensile or compressive member which consists of two or more materials in parallel is called a *compound bar*. Although this term is normally applied only to metals, it is extended in this chapter to include axially loaded reinforced concrete columns.

SPECIMEN QUESTION 8

A short, rigid bar is suspended from an indented rigid support by a central vertical steel rod and two symmetrically placed vertical copper rods. The rigid bar is loaded with a centrally placed "load," to give a total of 3 Mg. The area of the steel rod is 60 mm² and the area of each copper rod 150 mm²; the length of the copper rods is 3 metres and the length of the steel rod 3·6 metres. Young's modulus for steel = 205 kN/mm². Young's modulus for copper = 93 kN/mm². Determine (*a*) the stress in each rod, and (*b*) the extension of each rod.

SOLUTION

The system is shown in Fig. 12. Each rod will carry only a proportion of the 29·4 kN load. In a question with a compound bar, for each material there will be a value of stress, strain, load, cross-sectional area, original length and change in length. The point to look for in any such problem is, which of these is the same for both materials. In this case it is the change in length $\delta L$ which is common to both materials, e.g.:

for copper (one bar only):

$E_c = 93$ kN/mm$^2$; $L_c = 3$ m; $A_c = 150$ mm$^2$; $W_c = ?$

for steel:

$E_s = 205$ kN/mm$^2$; $L_s = 3\cdot6$ m; $A_s = 60$ mm$^2$; $W_s = ?$ $\left.\right\}\delta L_c = \delta L_s = ?$

$\qquad$ also $2W_c + W_s = 29\cdot4$ kN

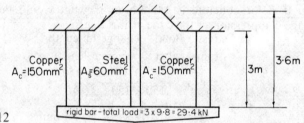

FIG. 12

$$E = \frac{\text{stress}}{\text{strain}} = \frac{\text{load}}{\text{area}} \times \frac{\text{original length}}{\text{change in length}}$$

for copper:

$$93 = \frac{W_c}{150} \times \frac{3000}{\delta L_c} \quad \text{or} \quad \delta L_c = 0\cdot215\ W_c$$

for steel:

$$205 = \frac{W_s}{60} \times \frac{3600}{\delta L_s} \quad \text{or} \quad \delta L_s = 0\cdot293\ W_s$$

but $\qquad\qquad\qquad \delta L_c = \delta L_s$

$\therefore \qquad\qquad\qquad W_c = \dfrac{0\cdot293}{0\cdot215}\ W_s$

or $\qquad\qquad\qquad W_c = 1\cdot36\ W_s$

also $\qquad\qquad\qquad 2W_c + W_s = 29\cdot4$

$\qquad\qquad\qquad (2 \times 1\cdot36 + 1)\ W_s = 29\cdot4$

$\qquad\qquad\qquad W_s = \dfrac{29\cdot4}{3\cdot72} = 7\cdot9$ kN

and $\qquad\qquad\qquad W_c = 1\cdot36 \times 7\cdot9 = 10\cdot75$ kN

∴ Stress in copper $f_c = 10\,750/150 = \underline{\underline{72\ \text{N/mm}^2}}$

Stress in steel $f_s\quad = 7900/60 = \underline{\underline{132\ \text{N/mm}^2}}$

Extension $\delta L_c = \delta L_s = 0.215 \times 10.75 = \underline{\underline{2.31\ \text{mm}}}$

SPECIMEN QUESTION 9

A solid circular rod of mild steel and a similar rod of brass are securely joined together in line at point C and are heated uniformly. Each rod is 650 mm² in cross-sectional area.

The two free ends of the rods are rigidly clamped at A and B as shown in Fig. 13, such that distance AC is 750 mm and CB 600 mm. Calculate (a) the force in the rods when they have cooled 41 °C below initial temperature and (b) the new distance AC at this time.

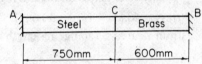

FIG. 13

Modulus of elasticity of mild steel = 205 kN/mm²
Modulus of elasticity of brass = 91 kN/mm²
Coefficient of linear expansion of steel = 0.0000162 per °C
Coefficient of linear expansion of brass = 0.0000261 per °C

SOLUTION

In this problem the stresses are set up by the rigid supports A and B resisting contraction due to fall in temperature.

The total contraction due to temperature must equal the total expansion due to stress, or:

$$L_s\alpha_s T + L_b\alpha_b T = \frac{F_s}{A_s} \times \frac{L_s}{E_s} + \frac{F_b}{A_b} \times \frac{L_b}{E_b}$$

$750 \times (16.2 \times 10^{-6}) \times 41$
$\quad + 600 \times (26.1 \times 10^{-6}) \times 41 = \dfrac{F_s}{650} \times \dfrac{750}{205} + \dfrac{F_b}{650} \times \dfrac{600}{91}$

$$0.498 + 0.642 = 0.0056F_s + 0.010F_b$$

Also the force in each rod caused by the supports must be the same, i.e. $F_s = F_b$

∴ $\qquad F_s = F_b = 1.14/0.0156 = 73.1\ \text{kN}$

Contraction of steel due to temperature = 0.498 mm
Expansion of steel due to force = $0.0056 \times 73.1 = 0.409$ mm
Actual change in length of steel = $0.498 - 0.409$
$\qquad\qquad\qquad\qquad\qquad = 0.089$ mm contraction.
Final length of AC = $750 - 0.089 = \underline{\underline{749.911\ \text{mm}}}$

SPECIMEN QUESTION 10

A short reinforced concrete column 250 mm square carries an axial "load" of 30 Mg. The column is reinforced with four 40 mm diameter bars. If the modulus of elasticity for steel is fifteen times that for concrete, find the stresses in the steel and the concrete. If the concrete stress must not exceed $3.9$ N/mm$^2$ what area of steel is required for the column to support 60 Mg?

SOLUTION

This problem concerns the design of a reinforced concrete column. The ratio $E_s/E_c = 15$ is known as the modular ratio $m$ and is used in reinforced concrete design.

The steel and concrete are bonded together and therefore they must shorten by equal amounts, hence both suffering equal strain.

$$E = \frac{\text{stress}}{\text{strain}} = \frac{\text{load}}{\text{area}} = \frac{\text{original length}}{\text{change in length}}$$

For steel:

$$\text{area of steel} = 4 \times (\pi \times 40^2/4) = 5030 \text{ mm}^2$$
$$E_s = (W_s/5030) \div \text{strain in steel} \tag{1}$$

For concrete:

$$\text{area of concrete} = 250 \times 250 - 5030 = 57\,470 \text{ mm}^2$$
$$E_c = (W_c/57\,470) \div \text{strain in concrete} \tag{2}$$

But $E_s/E_c = 15$, and strain in steel = strain in concrete. From (1) and (2):

$$\frac{E_s}{E_c} = \frac{W_s}{5030} \times \frac{57\,470}{W_c} = 15$$

or

$$\frac{W_s}{W_c} = \frac{15 \times 5030}{57\,470} = 1.3$$

also

$$W_s + W_c = 30 \times 9.8 = 294$$

$\therefore$

$$1.3\,W_c + W_c = 294$$

$$W_c = \frac{294}{2.3} = 128 \text{ kN}$$

$$W_s = 294 - 128 = 166 \text{ kN}$$

Stress in steel $f_s = 166\,000/5030 = \underline{33 \text{ N/mm}^2}$

Stress in concrete $f_c = 128\,000/57\,470 = \underline{\underline{2.23 \text{ N/mm}^2}}$

For $60 \times 9.8 = 588$ kN load, let $A_s$ = area of steel required

For steel:

$$E_s = (W_s/A_s) \div \text{strain in steel} \tag{3}$$

For concrete:

$$E_c = \frac{W_c}{(62\,500 - A_s)} \div \text{strain in concrete} \tag{4}$$

Also
$$E_s/E_c = 15 \tag{5}$$
$$W_c + W_s = 588\,000 \tag{6}$$
$$\frac{W_c}{(62\,500 - A_s)} = 3{\cdot}9 \tag{7}$$

From (7):

$$W_c = 3{\cdot}9(62\,500 - A_s) = 244\,000 - 3{\cdot}9A_s$$

Substitute in (6):

$$244\,000 - 3{\cdot}9A_s + W_s = 588\,000$$
$$W_s = 344\,000 + 3{\cdot}9A_s$$

From (3) (4) and (5) (since strain in concrete = strain in steel):

$$\frac{E_s}{E_c} = \frac{W_s}{A_s} \times \frac{(62\,500 - A_s)}{W_c} = 15$$

or
$$15 = \frac{(344\,000 + 3{\cdot}9A_s)}{A_s} \times \frac{(62\,500 - A_s)}{3{\cdot}9(62\,500 - A_s)}$$

$$\therefore \qquad 58{\cdot}5A_s = 344\,000 + 3{\cdot}9A_s$$
$$A_s = 344\,000/54{\cdot}6 \qquad = \underline{\underline{6300 \text{ mm}^2}}$$

or four bars of 45 mm diameter.

$$\text{Total area} = (4 \times \pi \times 45^2)/4 \qquad = 6362 \text{ mm}^2$$

(It is normal to use slightly more than the economical amount of steel using nominal size bars.)

**Transverse strain**
When a bar is loaded in tension it increases its length but decreases in cross-sectional dimensions. Similarly a compressive load produces a decrease in length but an increase in cross-sectional dimensions.

The transverse strain is defined as the ratio of change in width to original width. Within the limit of proportionality, transverse strain varies as longitudinal strain, or:

$$\frac{\text{transverse strain}}{\text{longitudinal strain}} = \text{constant } (\sigma)$$

This constant $\sigma$ is known as Poisson's ratio, and its value normally lies between $\frac{1}{3}$ and $\frac{1}{4}$.

## Modulus of rigidity

When shear stress only is being considered the ratio of shear stress to shear strain is known as the modulus of rigidity, i.e.:

$$\text{modulus of rigidity } G = f_s/\phi$$

It can be shown that the relationship between modulus of elasticity and modulus of rigidity is $E = 2G(1 + \sigma)$.

SPECIMEN QUESTION 11

In a tensile test on a steel tube of 20 mm external diameter and 12 mm internal diameter, it was found that the increase in length of 200 mm gauge length was 0·0474 mm per Mg and the decrease in external diameter was 0·00133 mm per Mg. Calculate Young's modulus and Poisson's ratio and deduce the modulus of rigidity.

SOLUTION

Original cross-sectional area of metal $= \pi(20^2 - 12^2)/4 = 201$ mm$^2$
For 1 Mg $= 9\cdot8$ kN load, stress $= 9800/201 = 48\cdot8$ N/mm$^2$
           Longitudinal strain $= 0\cdot0474/200 = 2\cdot37 \times 10^{-4}$

$$\therefore \quad E = \frac{\text{stress}}{\text{strain}} = \frac{48\cdot7}{2\cdot37 \times 10^{-4}}{}^1 \qquad = 205\,000 \ \pi/\text{Nm}^2$$

$$\text{Lateral strain} = \frac{(\text{decrease in diameter})}{(\text{original diameter})} = \frac{1\cdot33 \times 10^{-4}}{20}$$

$$= 6\cdot65 \times 10^{-5}$$

Poisson's ratio $\sigma = 6\cdot65 \times 10^{-5}/2\cdot37 \times 10^{-4} = 0\cdot28$

$$E = 2G(1 + \sigma)$$
$$G = \frac{20\,500}{2(1 + 0\cdot28)}$$
$$= 8008 \ \text{N/mm}^2$$

## Plasticity

In this chapter only the elastic portion of the stress–strain graph has been considered. In mild steel the plastic portion shows strains up to ten times as great, at a constant maximum stress. Modern design methods deal with this part of the curve but this is beyond the scope of this volume.

EXAMINATION QUESTIONS

1. The following is a set of results obtained from an experiment to determine the 0·1% proof stress of a specimen of duralumin, 15 mm × 6 mm cross-section and 100 mm gauge length.

| "Load" (Mg): | 0·25 | 0·5 | 0·75 | 1·0 | 1·25 | 1·5 | 1·75 |
|---|---|---|---|---|---|---|---|
| Extension (mm): | 0·033 | 0·074 | 0·112 | 0·151 | 0·192 | 0·230 | 0·272 |

"Load" (Mg): 2·0  2·25  2·5  2·75  3·0  3·25  3·26
Extension (mm): 0·312 0·350 0·400 0·450 0·521 0·620 0·950

"Load" (Mg): 3·27
Extension (mm): Failed

Determine the value of the proof stress, and the elastic modulus for the specimen.

2. A steel rod of cross-sectional area 1000 mm² and a co-axial copper tube of cross-sectional area 1600 mm², are firmly attached at their ends to form a compound bar. Determine the stress in the steel and in the copper when the temperature of the compound bar is raised by 80 °C and an axial tensile "force" of 6 Mg is applied at the ends.

The moduli of elasticity for steel and copper are 205 kN/mm² and 105 kN/mm² and the coefficients of linear expansion are 11 × 10⁻⁶ per °C and 16·5 × 10⁻⁶ per °C respectively.

3. Two supports are propped apart by the compound bar shown in Fig. 14. Part AB is steel of uniform cross-sectional area 5000 mm² and part BC is aluminium of uniform cross-sectional area 2000 mm². The whole assembly is stress free at a temperature of 16 °C. The temperature rises to 32 °C and during the process the right hand support C yields a distance of 0·025 mm. Calculate the stress then existing in the aluminium.

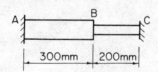

FIG. 14

The moduli of elasticity for steel and aluminium are 205 kN/mm² and 68·5 kN/mm² and the coefficients of linear expansion are 11·7 × 10⁻⁶ per °C and 23 × 10⁻⁶ per °C respectively.

4. A reinforced concrete column is required to carry an axial "load" of 66 Mg. Assuming the column to have four reinforcing bars and taking the maximum allowable stress in the concrete as 7·4 N/mm² and the modular ratio $m$ as 15, determine suitable dimensions for the section and reinforcing bars:

(a) for a square section with 1% area of steel;
(b) for a rectangular section b × 1·5b with four 16 mm diameter bars.

5. A steel bar 30 mm in diameter was used in a tension test, when it was found that an axial "load" of 10 Mg produced an elastic

extension of 0·153 mm and a reduction in diameter of 0·0055 mm on a gauge length of 230 mm.

If a bolt of the same material and diameter is used to connect two plates subject to a tensile "load" of 7·5 Mg, estimate the shear strain in the bolt.

## CHAPTER 2

# STRUCTURES SUBJECT TO DIRECT STRESS

THE *resultant* of a system of co-planar forces is the single force which could replace all the forces in the system. The *equilibrant* is equal in magnitude but opposite in direction to the resultant, and would maintain the system in a static position.

## RESOLUTION OF FORCES

Any single force can be resolved into two component forces acting in any two directions at right angles to each other, e.g. find the vertical and horizontal components of force $F$ acting at angle $\theta$ to the horizontal. This is shown in Fig. 15(a); Fig. 15(b) shows the solution, using a simple triangle of forces.

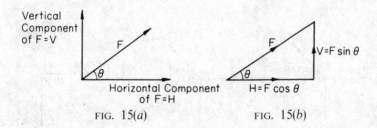

FIG. 15(a)                    FIG. 15(b)

To draw the triangle of forces, the components and the force are drawn parallel to their original directions to form a triangle. The solution of the components is then a matter of simple trigonometry. Alternatively, if the two components $V$ and $H$ are given, the resultant $F$ may be found, i.e. $F = \sqrt{(V^2 + H^2)}$.

This resolution of forces can be used to find the resultant (or equilibrant) of a system of forces.

SPECIMEN QUESTION 12
Calculate the magnitude of the resultant of the system of loads shown in Fig. 16(a) and find its angle of inclination to the horizontal.

SOLUTION
Figure 16(b) shows the conversion to force units.
    First find the vertical components of all the forces and add them algebraically.

19

$$\Sigma V\uparrow = 4 \sin 45° - 3 \sin 30° - 2$$
$$[\sin 45° = 1/\sqrt{2}, \quad \sin 30° = 0\cdot5]$$
$$\therefore \quad \Sigma V\uparrow = 4/\sqrt{2} - 3/2 - 2 = 2\cdot83 - 1\cdot5 - 2 = -0\cdot67 \text{ kN}$$

The minus sign shows that the wrong direction was chosen for $\Sigma V$ (upwards ↑), i.e. the total vertical component is actually 0·67 kN downwards.

Next find the horizontal components of all the forces and add them algebraically:

$$\underset{\leftarrow}{\Sigma H} = 6 - 4 \cos 45° - 3 \cos 30°$$
$$[\cos 45° = 1/\sqrt{2}, \quad \cos 30° = (\sqrt{3})/2]$$
$$\therefore \quad \underset{\leftarrow}{\Sigma H} = 6 - 4/\sqrt{2} - 3(\sqrt{3})/2 = 6 - 2\cdot83 - 2\cdot6 = 0\cdot57 \text{ kN}$$

In this case the correct direction was chosen (to the left ←) and $\Sigma H$ is 0·57 kN to the left.

It should also be noted that as $\cos 90° = 0$ any force at right angles to the direction of resolution has no effect on the component in that direction.

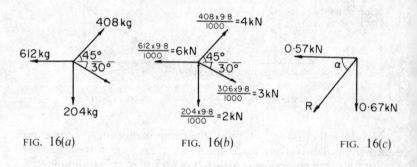

FIG. 16(*a*)           FIG. 16(*b*)           FIG. 16(*c*)

The summation of these two components is shown in Fig. 16(*c*).

The resultant of these two forces, $R = \sqrt{(0\cdot57^2 + 0\cdot67^2)} = 0\cdot88$ kN
Angle of resultant with horizontal $= \tan^{-1}\alpha$
$$\text{Tan } \alpha = V/H = 0\cdot67/0\cdot57 = 1\cdot175$$
$$\therefore \qquad\qquad\qquad \alpha = 49° \ 36'$$

Therefore the resultant of the system of forces shown in Fig. 16(*b*) is 0·88 kN acting to the left and inclined downwards at an angle of 49° 36' to the horizontal.

The equilibrant would be 0·88 kN acting to the right and inclined upwards at an angle of 49° 36' to the horizontal.

## CONDITIONS OF EQUILIBRIUM (Co-planar forces)

*In any structure it is essential that all forces be in equilibrium—* otherwise the structure would move! There are two basic conditions of equilibrium which are extremely important, and should be remembered at all times:

1. The algebraic sum of the components of all forces in any two directions at right angles to each other must each equal zero.
2. The sum of the moments of the forces about any point must equal zero.

These may be shown in symbols as:

$$\Sigma V = 0 \qquad \Sigma H = 0 \qquad \Sigma M = 0$$

*These three form the basis of all structural analysis.*

## MOMENTS

The moment of a force about a point is the product of the force and the shortest distance of the line of action of the force from the point

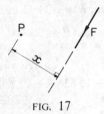

FIG. 17

(*see* Fig. 17). The moment of force $F$ about point $P = Fx$. This moment will have a clockwise rotating effect upon P, which, for equilibrium, would have to be balanced.

The units of moment are force × length, i.e. N m.

SPECIMEN QUESTION 13
The positions of three co-planar loads are fixed by reference to a vertical line XY as shown in Fig. 18(*a*). Calculate the magnitudes and directions of a horizontal force at X and an oblique force through Y such that the whole system is in equilibrium.

SOLUTION
Fig. 18(*b*) shows the final system of forces in equilibrium. The moments about any point must be checked. Point Y is chosen since the force $F_Y$ which is unknown in magnitude and direction passes through Y and hence will have no moment about Y. The shortest

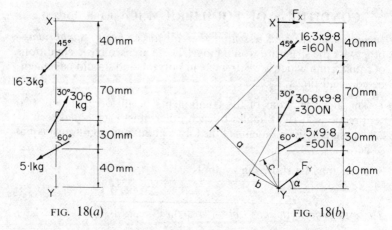

FIG. 18(*a*)                    FIG. 18(*b*)

distance from the line of action of each force to point Y must be calculated from Fig. 18(*b*):

$$a = (70 + 30 + 40) \sin 45 = 98.98 \text{ mm}$$
$$b = (30 + 40) \sin 30° = 35 \text{ mm}$$
$$c = 40 \sin 60° = 34.64 \text{ mm}$$

For equilibrium, $\Sigma V = 0$; $\Sigma H = 0$; $\Sigma M = 0$. Therefore about Y (clockwise positive $\circlearrowright + Ve$):

$$\Sigma M = 0 = 180F_X - 160 \times 98.98 + 300 \times 35 - 50 \times 34.64$$
$$0 = 180F_X - 15836.8 + 10500 - 1732$$
$$\therefore \quad F_X = +\frac{7068.8}{180} = +\underline{39.27 \text{ N}} \quad \text{(to the right, as shown)}.$$

Also the horizontal component of $F_Y = F_Y \cos \alpha$
and $\Sigma H = 0 = F_Y \cos \alpha + 50 \cos 30° - 300 \cos 60°$
$$+ 160 \cos 45° - 39.27$$

or $F_Y \cos \alpha = -50 \times (\sqrt{3})/2 + 300 \times 0.5 - 160 \times 1/\sqrt{2} + 39.27$

$$= -43.30 + 150 - 113.14 + 39.27$$
$$= 32.83 \text{ N} \quad \text{to the left.}$$

The vertical component of $F_Y = F_Y \sin \alpha$

and $\Sigma V = 0 = \downarrow F_Y \sin \alpha + 50 \cos 60° - 300 \cos 30° + 160 \cos 45°$

or $\downarrow F_Y \sin \alpha = -50 \times 0.5 + 300 \times (\sqrt{3})/2 - 160 \times 1/\sqrt{2}$

$$= -25 + 259.8 - 113.14$$
$$= 121.66 \text{ N} \quad \text{downwards.}$$

But   $(F_Y \sin \alpha)^2 + (F_Y \cos \alpha)^2 = F_Y^2 (\sin^2 \alpha + \cos^2 \alpha) = F_Y^2$

$$F_Y = \sqrt{(121 \cdot 66^2 + 32 \cdot 83^2)} = \underline{126 \text{ N}} \checkmark$$

$(F_Y \sin \alpha / F_Y \cos \alpha) = \tan \alpha = 121 \cdot 66/32 \cdot 83 = 3 \cdot 70$

$$\alpha = \underline{\underline{74° \ 53'}}$$

A useful exercise at this stage would be to take moments about other points to see that $\Sigma M = 0$, e.g. moments about X.

The vertical component of $F_Y$ passes through point X and therefore has no moment about X. The horizontal component of $F_Y = 126 \cos 74° \ 53' = 32 \cdot 86$ N.

∴        $M_X = 32 \cdot 86 \times 180 + 50 \cos 30° \times 140 - 300 \cos 60° \times 110 \cdot$
$$+ 160 \cos 45° \times 40$$
$$= 5910 + 6060 - 16\,500 + 4530 = 0$$

Moments about any other point will also give $\Sigma M = 0$

A common application of this theory is in the determination of reactions in a structure, e.g. determine the reactions at A and B for the simply supported beam loaded as shown in Fig. 19(a).

Figure 19(b) shows the system in force units.

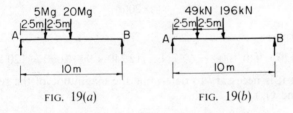

FIG. 19(a)                    FIG. 19(b)

Take moments about one of the unknown reactions (A).

$$49 \times 2 \cdot 5 + 196 \times 5 - R_B \times 10 = 0$$
$$R_B = (122 \cdot 5 + 980)/10 = \underline{110 \cdot 25 \text{ kN}}$$

Resolve vertically:

$$R_A = 196 + 49 - 110 \cdot 25 = \underline{134 \cdot 75 \text{ kN}}$$

This is normally the first step in the analysis of any simple structure.

## FORCES IN FRAMES

Many structures are made up into frames, such as roof trusses and Warren girders. Each member of such a frame is subject to stress and strain as discussed in Chapter 1. To determine the stress in each member, it is first necessary to calculate the force, and this can be done by a series of resolution of forces. The two main analytical methods of calculating the forces in a frame are shown in specimen questions 14 and 15.

### Resolution of joints

SPECIMEN QUESTION 14

A simply supported girder AG carries loads as shown in Fig. 20(*a*). Determine the forces in each member of the frame and state whether the member is in tension or compression. (All sloping members are at 60° to the horizontal.)

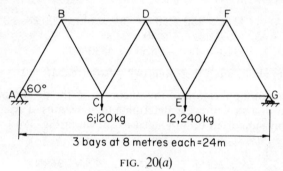

FIG. 20(*a*)

SOLUTION

$$(6120 \times 9.8)/1000 = 60 \text{ kN}; \quad (12\,240 \times 9.8)/1000 = 120 \text{ kN}.$$

First it is necessary to determine the magnitude of the reactions at A and G, i.e.:

$$R_A = (60 \times 16 + 120 \times 8)/24 = \quad 80 \text{ kN}$$
$$R_G = 60 + 120 - 80 \qquad\qquad = 100 \text{ kN}$$

Not only must the complete structure be in equilibrium, but so also must any part of the structure. For example, if joint A were "cut off" but the forces in members AB and AC maintained, then this portion must be in equilibrium ($\Sigma M = 0$, $\Sigma V = 0$, $\Sigma H = 0$).

Consider joint A (Fig. 20(*b*)). By inspection, for equilibrium there must be a downward component to balance the 80 kN. This is supplied by member AB, hence the direction of this force can be indicated. But $F_{AB}$ will now have a horizontal component to the left which must be balanced by $F_{AC}$. Hence the direction of $F_{AC}$ can be shown.

For equilibrium, $\Sigma V = 0$

$$\therefore \quad F_{AB} \sin 60° = 80 \qquad\qquad (\sin 60° = \sqrt{3}/2)$$
$$F_{AB} = 80 \times 2/\sqrt{3} = \underline{160/\sqrt{3} \text{ kN}}$$

Also         $\Sigma H = 0$

$\therefore$         $F_{AC} = F_{AB} \cos 60°$         (cos 60° = 0·5)
             $= 160/\sqrt{3} \times 0·5$      $= \underline{\underline{80/\sqrt{3} \text{ kN}}}$

*Note:* $\Sigma M$ will equal zero, since all the forces pass through the same point. The $\sqrt{3}$ denominator has been retained for simplicity of calculation.

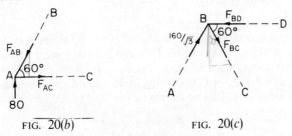

FIG. 20(*b*)                FIG. 20(*c*)

If member AB has a force of $160/\sqrt{3}$ kN ↙ at A, then for this member to be in equilibrium there must be a force in AB of $160/\sqrt{3}$ kN ↗ at B. Similarly, there must be a force of $80/\sqrt{3}$ kN at C. These forces are shown on the final diagram, Fig. 20(*n*).

Each joint in turn is now dealt with in a similar manner, selecting each time a joint with only two unknown forces.

Consider joint B (Fig. 20(*c*)). $F_{BC}$ must have a downward component, as shown, to balance the upward component of $F_{BA}$.

Resolving vertically, $F_{BC} \cos 30° = (160/\sqrt{3}) \cos 30°$,

$\therefore F_{BC} = \underline{\underline{160/\sqrt{3} \text{ kN}}}$

Resolving horizontally, $F_{BD} = (160/\sqrt{3}) \sin 30 + (160/\sqrt{3}) \sin 30$
$= \underline{\underline{160/\sqrt{3} \text{ kN}}}$. These are shown in Fig. 20(*n*).

Consider joint C (Fig. 20(*d*)). $F_{CD}$ must have an upward (or downward) component to balance $F_{CB}$ and the 60 kN load. Assume that $F_{CD}$ is upwards and resolve vertically.

$$F_{CD} \sin 60° = 60 - (160/\sqrt{3}) \sin 60°$$
$$F_{CD} \times \sqrt{3}/2 = 60 - 160/\sqrt{3} \times (\sqrt{3})/2 = -20.$$

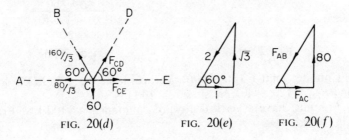

FIG. 20(*d*)           FIG. 20(*e*)              FIG. 20(*f*)

The minus sign shows that $F_{CD}$ at joint C is actually downwards.

$\therefore F_{CD} = 40/\sqrt{3}$ kN downwards.

This is shown correctly in Fig. 20(n).

By inspection, $F_{CE}$ must be to the right. Resolving horizontally,

$$F_{CE} = 80/\sqrt{3} + (160/\sqrt{3}) \cos 60° + (40/\sqrt{3}) \cos 60°$$
$$= 80/\sqrt{3} + 160/\sqrt{3} \times 0.5 + 40/\sqrt{3} \times 0.5 = \underline{180/\sqrt{3} \text{ kN}}$$

This is shown in Fig. 20(n).

The resolution of these joints can often be simplified, and with practice it should be possible to solve simple frames by inspection only. For instance, in this frame the basic triangle of forces would be as shown in Fig. 20(e). Considering joint A again, the triangle of forces for this joint would be as shown in Fig. 20 (f). By similar triangles:

$$F_{AB}/80 = 2/\sqrt{3}$$
$$\therefore \qquad F_{AB} = 160/\sqrt{3} \text{ kN}$$
$$F_{AC}/80 = 1/\sqrt{3}$$
$$\therefore \qquad F_{AC} = 80/\sqrt{3} \text{ kN}$$

The remainder of specimen question 14 will now be completed by quicker methods. It would be a useful exercise, however, to complete the question by formal resolution of joints, as a check, *always bearing in mind that the quick methods shown are still basically resolution of joints.*

Consider joint D (Fig. 20(g)). By inspection, $F_{DE}$ must be $40/\sqrt{3}$ kN $\searrow$ (to balance the vertical component of $F_{DC} \nearrow$).

Total horizontal force due to $F_{DC}$, $F_{DE}$ and $F_{DB}$

$$= (40/\sqrt{3}) \cos 60° + (40/\sqrt{3}) \cos 60°$$
$$+ 160/\sqrt{3} = 200/\sqrt{3}$$

Hence $\qquad F_{DF} = \underline{200/\sqrt{3} \text{ kN}}$

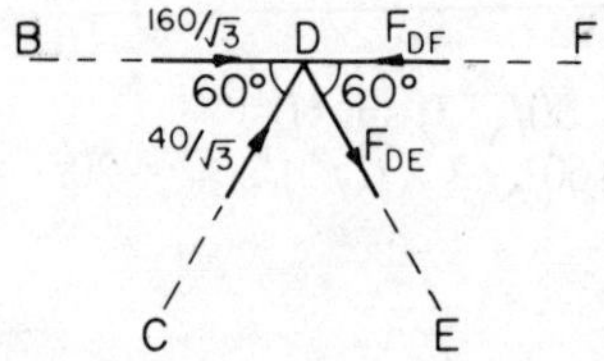

FIG. 20(g) 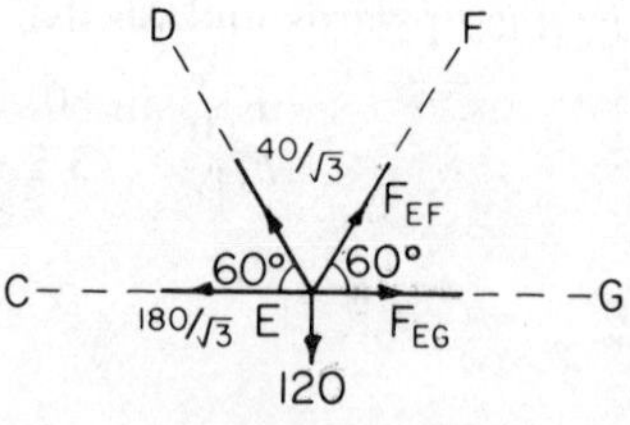          FIG. 20(h)

Consider joint E (Fig. 20(h)). Vertical force due to 120 kN load and $F_{ED} = 120 - (40/\sqrt{3}) \sin 60° = 100$ kN

$\therefore F_{EF}$ must have a vertical upward component of 100 kN. From Fig. 20(e),

$$100/F_{EF} = (\sqrt{3})/2$$
$$\therefore \qquad F_{EF} = \underline{200/\sqrt{3} \text{ kN}}$$

Horizontal force due to $F_{EC}$, $F_{ED}$ and $F_{EF} = 180/\sqrt{3} + (40/\sqrt{3})$ cos $60° - (200/\sqrt{3})$ cos $60°$

$$= 180/\sqrt{3} + (40/\sqrt{3}) \times 0\cdot5 - (200/\sqrt{3}) \times 0\cdot5 = 10/\sqrt{3}$$
$$\therefore \quad F_{EG} = \underline{100/\sqrt{3} \text{ kN}}$$

Consider joint F (Fig. 20($j$)). By inspection, three forces each at 120° meeting at a point must all be equal, $\therefore$ $F_{FG} = 200/\sqrt{3}$ kN. This gives the force in all members of the frame. It would be a useful check to ensure that joint G *is* in equilibrium. Consider joint G

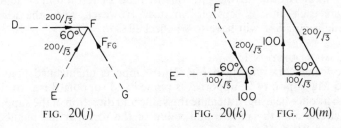

FIG. 20($j$)  FIG. 20($k$)  FIG. 20($m$)

(Fig. 20($k$)), which would give the triangle of forces shown in Fig. 20($m$). Compare this triangle with figure 20($e$), and if joint G is in equilibrium, then:

$$\frac{200/\sqrt{3}}{2} = \frac{100/\sqrt{3}}{1} = \frac{100}{\sqrt{3}}$$

which is correct. Therefore joint G is in equilibrium.

The final solution to the problem is shown in Fig. 20($n$). *Note:* the $\sqrt{3}$ denominator has been retained for simplicity of calculation.

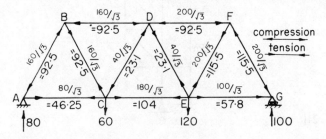

all forces in kilonewtons

FIG. 20($n$)

A great deal of space has been given to the solution of this question to bring out various important points. With practice, it should be possible to solve simple frames such as this with no written working at all. Questions 2, 3 and 4 at the end of this chapter are further examples, and should now be attempted.

An assumption in this type of question is that all joints can rotate freely. This assumption is rarely absolutely correct, but if the joints are not too rigid it gives a reasonable solution.

## Method of sections

If the forces in only a few members are required, then it may be quicker to use the method of sections. In this method, the whole frame is "cut" and one part only checked for equilibrium. This is in fact what is done in the previous method of resolution of joints, where each joint is "cut off" in turn. However, when the whole frame is cut, $\Sigma M$ will have to be checked.

SPECIMEN QUESTION 15

The framework shown in Fig. 21(a) supports an inclined load of 24·5 Mg at point C. Support A is pinned and support B is provided with a roller bearing. Calculate the value and direction of the support reactions and the nature and value of the force in the members marked W, X, Y and Z.

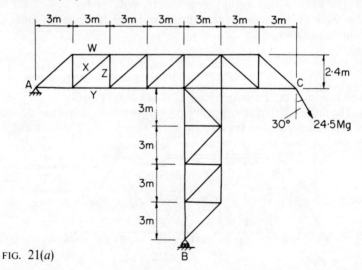

FIG. 21(a)

SOLUTION

$$\text{Force at C} = 24·5 \times 9·8 = 240 \text{ kN}$$

The reaction at B must be vertical, as any horizontal component would cause the rollers to move. The reaction at A must therefore have horizontal and vertical components ($H_A$ and $V_A$ respectively).

Take moments about A (hence excluding reactions at A and also horizontal components of 240 kN):

$$240 \cos 30° \times 21 - V_B \times 12 = 0$$
$$V_B = 420 \cos 30° = 420(\sqrt{3})/2 = \underline{\underline{364 \text{ kN}}} \quad \text{(upwards)}.$$

Resolve vertically:

$$\downarrow V_A + 240 \cos 30° - 364 = 0$$
$$\downarrow V_A = 364 - 208 = \underline{\underline{156 \text{ kN}}} \quad \text{(downwards)}.$$

Resolve horizontally:

$$H_A - 240 \sin 30° = 0$$

$$\overline{H}_A = 240 \times 0.5 = \underline{\underline{120 \text{ kN}}} \quad \text{(to the left)}.$$

$$R_A = \sqrt{(120^2 + 156^2)} = \underline{\underline{197 \text{ kN}}}$$

If $R_A$ acts at $\alpha$ to the horizontal, $\tan \alpha = 156/120 = 1.3$ and $\alpha = \underline{\underline{52° \ 30'}}$

$\therefore$ *$R_A$ is 197 kN downwards to the left at 52° 30′ to the horizontal. $R_B$ is 364 kN vertically upwards.*

To determine the force in members W, X and Y, assume the frame is cut through these members and consider only the left-hand portion as in Fig. 21(*b*). This part of the frame must be in equilibrium

$$\Sigma V = 0 \quad \therefore F_X \cos \theta = V_A$$
but $$\cos \theta = \frac{2.4}{\sqrt{(3^2 + 2.4^2)}} = \frac{2.4}{3.842} = 0.625$$
$\therefore$ $$F_X = 156/0.625 = 250 \text{ kN} \nearrow$$

But as member X must be in equilibrium, there is a reaction at the other end; hence $\quad F_X = \underline{\underline{250 \text{ kN tension}}} \nearrow$

Take moments about P. $\Sigma M = 0$.

$\therefore$ $$F_W \times 2.4 = 3 \times V_A$$
$$F_W = (3 \times 156)/2.4 = \underline{\underline{195 \text{ kN tension}}} \leftarrowtail$$

Also, $\Sigma H = 0$.

$\therefore$

$$F_Y = H_A - F_W - F_X \sin \theta$$
$$= 120 - 195 - 250 \times 3/3.842$$
$$= 120 - 195 - 195 = -270$$
$$\underline{\underline{\text{or } 270 \text{ kN compression} \leftarrow\rightarrow}}$$

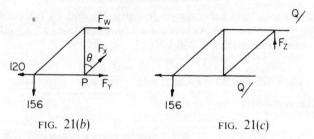

FIG. 21(b)          FIG. 21(c)

To find the force in member Z, cut the frame along QQ, as shown in Fig. 21(c).

$$\Sigma V = 0, \quad \therefore F_Z = \underline{156 \text{ kN compression } \updownarrow}$$

In some problems, a combination of these two methods may be employed.

SPECIMEN QUESTION 16

The plane frame shown in Fig. 22(a) is pin-jointed at A, B, C, D, E and F. Determine the forces in all members under the given loading. (Members AE and DF are not connected where they cross other members.)

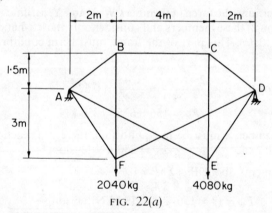

FIG. 22(a)

SOLUTION

(2040/1000) Mg × 9·8 = 20 kN;   (4080/1000) Mg × 9·8 = 40 kN

$$R_A = \frac{20 \times 6 + 40 \times 2}{8} = 25 \text{ kN}$$

$$R_D = 20 + 40 - 25 = 35 \text{ kN}$$

Consider a vertical cut between B and C, as shown in Fig. 22(b) and the forces acting in the directions shown.

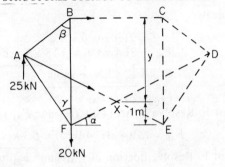

FIG. 22(b)

*Note:*         $y = 1\cdot5 + 3 - 1 = 3\cdot5$ m (*see* Fig. 22(b))
$\text{Sin } \alpha = 3/\sqrt{(3^2 + 6^2)} = 0\cdot447$
$\text{Cos } \alpha = 6/\sqrt{(3^2 + 6^2)} = 0\cdot894$
$\text{Sin } \beta = 2/\sqrt{(1\cdot5^2 + 2^2)} = 0\cdot8$
$\text{Cos } \beta = 1\cdot5/\sqrt{(1\cdot5^2 + 2^2)} = 0\cdot6$
$\text{Sin } \gamma = 2/\sqrt{(3^2 + 2^2)} = 0\cdot55$

Take moments about X (hence eliminating $F_{AE}$ and $F_{FD}$):

$$25 \times 4 - 20 \times 2 + F_{BC} \times 3\cdot5 = 0$$
$$F_{BC} = -60/3\cdot5 = -17\cdot1 \text{ (opposite to direction shown),}$$
or 17·1 kN $\longleftrightarrow$ compression.

Resolve vertically:

$$25 - 20 + F_{FD} \sin \alpha - F_{AE} \sin \alpha = 0$$
$$F_{AE} - F_{FD} = 5/0\cdot447 = 11\cdot2 \qquad (1)$$

Resolve horizontally:

$$F_{BC} + F_{AE} \cos \alpha + F_{FD} \cos \alpha = 0$$
$$F_{AE} + F_{FD} = 17\cdot1/0\cdot894 = 19\cdot1 \qquad (2)$$

Adding (1) and (2):

$$2F_{AE} = 30\cdot4$$
$$F_{AE} = 15\cdot2 \text{ kN} \rightarrow \text{tension.}$$
$$F_{FD} = 19\cdot1 - 15\cdot2 = 3\cdot9 \text{ kN} \rightarrow \text{tension.}$$

Now, using the method of resolution of joints, consider joint B. Resolve horizontally:

$$17\cdot1 = F_{AB} \sin \beta$$
$$F_{AB} = 17\cdot1/0\cdot8 = 21\cdot4 \text{ kN compression.}$$

Similarly,         $F_{CD} = 21\cdot4$ kN compression.

Resolve vertically:

$$21\cdot4 \cos \beta = F_{BF}$$
$$F_{BF} = 21\cdot4 \times 0\cdot6 = \underline{\underline{12\cdot8 \text{ kN tension.}}}$$

Similarly,                $F_{CE} = \underline{\underline{12\cdot8 \text{ kN tension.}}}$

$$F_{FD} \cos \alpha = F_{AF} \sin \gamma$$

Consider joint F. Resolve horizontally (assume $F_{AF}$ to the left):

$$F_{AF} = (3\cdot9 \times 0\cdot894/0\cdot55) = \underline{\underline{6\cdot3 \text{ kN tension.}}}$$

Consider joint E. Resolve horizontally (assume $F_{ED}$ to the right):

$$F_{AE} \cos \alpha = F_{ED} \sin \gamma$$
$$F_{ED} = (15\cdot2 \times 0\cdot894)/0\cdot55 \qquad = \underline{\underline{24\cdot7 \text{ kN tension.}}}$$

### Graphical solution

The forces in a frame may also be solved by graphical means. Although commonly used in practice, this tends to be tedious in an examination problem. A simple example is given here to show the principles involved.

SPECIMEN QUESTION 17

The pin-jointed roof truss in Fig. 23(a) is subject to the loads shown in the diagram. Determine, either graphically or analytically, the forces in members W, X, Y, Z and state whether the forces are tensile or compressive.

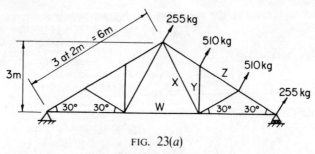

FIG. 23(a)

SOLUTION

$$(255/1000) \times 9\cdot8 = 2\cdot5 \text{ kN}; \quad (510/1000) \times 9\cdot8 = 5 \text{ kN}$$

The first step is to draw the truss to scale, as in Fig. 23(b) (1:200). The members to be analysed are shown by heavy lines. As it will be necessary to calculate the vertical reaction at the right hand support ($V_R$), dimensions PR and LR will be required. These could be scaled, but are easily calculated in this case (9 m and $6\sqrt{3}$ m).

To find $V_R$, take moments about L:

$$V_R \times 6\sqrt{3} = 2.5 \times 3 + 5 \times 5 + 5 \times 7 + 2.5 \times 9$$
$$V_R = (2.5/6\sqrt{3})(3 + 10 + 14 + 9) = 8.66 \text{ kN}$$

(downwards)

(this could also be found by graphical construction).

To construct the force diagram for this truss (Fig. 23(c)), Bowes' notation has been used, i.e. each space between forces (internal member or external load) is marked with a capital letter (Fig. 23(b)). When these forces are indicated on the force diagram (Fig. 23(c)) they are noted by the letters on each side of the force, but in lower case (small letters). Thus member X in the truss is between G and H and is indicated on the force diagram as gh. gh will be parallel to member X and its magnitude will be proportional to the force in member X. It is important always to give the scale of the diagrams.

The force diagram is constructed in the following stages.

(i) Choose a force, the direction and magnitude of which is known, and draw a line parallel to it and scaled to the same magnitude, i.e. the 2·5 kN force at the top of the truss which lies between the letters A and B. This is shown on the force diagram as line ab. Note that the direction ab is the same direction as the 2·5 kN force.

(ii) Choose another known force which lies adjacent to A or B and draw this to scale on the force diagram, i.e. upper 5 kN force lies between B and C and is shown on the force diagram as bc.

(iii) Continue from force to force as far as possible, i.e. a–b–c–d–e–f, the force ef being the vertical reaction at R.

(iv) The next external force is the reaction $R_L$, which is unknown in magnitude and direction. However, it lies between F and A. Therefore line fa on the force diagram can be scaled to give the magnitude and direction of $R_L$ (8·66 kN downwards to the left at 30° to the horizontal).

This completes the external force diagram (shown in heavy lines). All force diagrams for systems in equilibrium must close in the manner shown. Continue by constructing force diagrams for each of the joints in the frame.

(v) Select a joint with only two unknown forces, i.e. joint R.

(vi) On the force diagram, de is already drawn, as is ef. Draw a line parallel to FK (horizontal) through f. Point k must lie on this line.

(vii) To close the diagram, kd must be parallel to KD and can be drawn in only one place, hence locating point k.

FIG. 23(b)

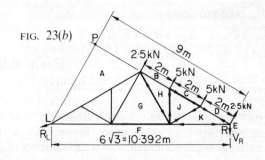

FIG. 23(c)

Scale 10mm≡2kN

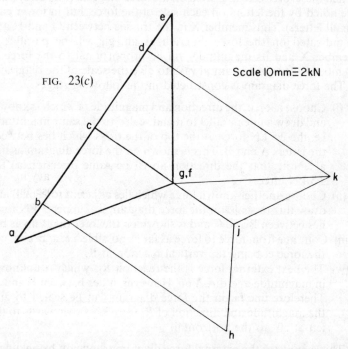

This closes the diagram for joint R, and the direction and magnitude of fd and kd can be scaled if required. The direction of fd and kd follow round the force diagram as before and should be marked on the *truss* at R, *not* on the force diagram.

For the joint CDKJ the force diagram will be cdkj, giving the forces at that joint.

Completion of joints HBCJ and GHJKF will give all the members required in this question.

Point g is found to coincide with point f, which means that there is no force in member GF.

The answers, scaled from Fig. 23(c), are:

$$F_W = hg = 0.$$
$$F_X = gh = 9{\cdot}8 \text{ kN compressive.}$$
$$F_Y = hj = 5{\cdot}64 \text{ kN tensile.}$$
$$F_Z = cj = 9{\cdot}9 \text{ kN tensile.}$$

For accuracy, the force diagram should be drawn to a large scale —at least twice as large as that of Fig. 23(c).

## BUCKLING OF COMPRESSION MEMBERS (Axial load)

If the force in any member of a frame is applied axially then that member is subject to direct stress. With tension forces the tensile stress is then simply load divided by area. With long slender members in compression, however, there is a tendency to buckle, and the maximum buckling load may be less than the maximum permissible load due to direct stress alone.

### Euler's theory for long struts

Euler's theory shows that for axially loaded, pin-ended struts the buckling load $P_e = (\pi^2 EI)/L^2$, where $E$ is the modulus of elasticity for the materials, $I$ is the least second moment of area about a central axis for the section, and $L$ is the effective length of the strut. The full derivation of this equation is beyond the scope of this volume. (See *Theory of Structures* by M. J. Smith and Brian J. Bell, in this series.)

### Second moment of area

The determination of the second moment of area about a central axis of any section is required to determine stress in any structural member subject to bending. The second moment of area about a central axis is equal in magnitude to the moment of inertia of the section and is commonly denoted by the letter $I$. Strictly speaking however, moments of inertia are used in dynamics and for full coverage any good textbook in applied mathematics should be referred to. As second moments of area for some simple sections are required in structural analysis, a few basic rules are given here.

1. For the second moment of area $I_c$ of a rectangle, breadth $b$, depth $d$, about a central axis (Fig. 24), $I_c = bd^3/12$.
2. For the second moment of area $I_X$ of any section of cross-sectional area $A$, about an axis $XX$, distance $y$ from the central axis and parallel to it (Fig. 25), $I_c = I_X - Ay^2$ or $I_X = I_c + Ay^2$.

For example to determine the second moment of area of a rectangle about its base (Fig. 26):

$$I_c = \frac{bd^3}{12}$$

$$A = bd \qquad y = d/2$$

$$\therefore \qquad I_x = I_c + Ay^2$$

$$= \frac{bd^3}{12} + bd \times \frac{d^2}{4}$$

$$= bd^3(\tfrac{1}{12} + \tfrac{1}{4}) = \frac{bd^3}{3}$$

which is another well-known value.

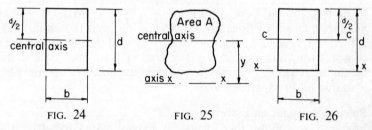

FIG. 24                    FIG. 25                    FIG. 26

3. The moment of inertia equals the area of section multiplied by the radius of gyration squared: $I = Ak^2$. The meaning of "radius of gyration" is again more the concern of dynamics than structures, but its magnitude is of value in determining the permissible stress in struts.

SPECIMEN QUESTION 18

The section shown in Fig. 27(a) is to be used as a strut of 7 m length. Determine the second moment of area of the section about each central axis and hence the slenderness ratio of the strut.

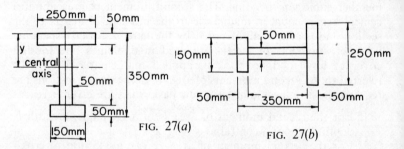

FIG. 27(a)

FIG. 27(b)

SOLUTION

The determination of $I$ for a built-up section is usually best determined by splitting the section into simple parts and summing up in a table, i.e.:

| 1 | 2 | 3 | 4 | 5 | 6 | 7 | 8 | 9 |
|---|---|---|---|---|---|---|---|---|
| Part of section | $b$ $(\times 10^{-1})$ | $d$ $(\times 10^{-1})$ | $A$ $(\times 10^{-2})$ | $\bar{x}$ $(\times 10^{-1})$ | $A\bar{x}$ $(\times 10^{-3})$ | $I_s$ $(\times 10^{-4})$ | $\bar{y}$ $(\times 10^{-1})$ | $A\bar{y}^2$ $(\times 10^{-9})$ |
| — | 25 | 5 | 125 | 2·5 | 312·5 | 260·4 | 12·7 | 20 161·3 |
| l | 5 | 25 | 125 | 17·5 | 2187·5 | 6510·4 | 2·3 | 661·3 |
| — | 15 | 5 | 75 | 32·5 | 2437·5 | 156·3 | 17·3 | 22 446·8 |
|  |  |  | $\Sigma 325$ |  | $\Sigma 4937·5$ | $\Sigma 6927·1$ |  | $\Sigma 43\,269·4$ |

*Column 5:* $\bar{x}$ is the distance from the centroid of the part of section being considered to the top of the whole section. The depth of centroid of the whole section will be $(\Sigma$ column 6$)/(\Sigma$ column 4$)$, or depth of centroid $y = (4937·5/325) \times 10 = 152$ mm.

*Column 7:* $I_s$ is the $I$ about the centroid of the part of section being considered.

*Column 8:* $\bar{y}$ is the distance from the centroid of the part of section being considered to the centroid of the whole section.

The $I$ about the centroid:

1. For the top flange $(I_s + A\bar{y}^2) = (260·4 + 20\,161·3) \times 10^4 = 204·217 \times 10^6$
2. For the web $= (6510·4 + 661·3) \times 10^4 = 71·716 \times 10^6$
3. For the bottom flange $= (156·3 + 22\,446·8) \times 10^4 = 226·03 \times 10^6$

$$501·963 \times 10^6$$

The sum of these is the second moment of area of the section about a horizontal central axis. It would, of course, give the same answer if $\Sigma$ column 7 is added to $\Sigma$ column 9,

or for section, horizontal axis $I_C = 501·963 \times 10^6$ mm$^4$

The radius of gyration $k_H = \sqrt{(I/A)}$

$$= \sqrt{\left(\frac{501·963 \times 10^6}{325 \times 10^2}\right)}$$

$$= 124 \text{ mm}$$

The value of $I$ about a vertical axis is easily found, as it is the sum of the $I$ value for three rectangles (Fig. 27(*b*)).

Vertical axis $I_c = \left(\dfrac{5 \times 15^3}{12} + \dfrac{25 \times 5^3}{12} + \dfrac{5 \times 25^3}{12}\right) \times 10^4$

$$= (1406·3 + 260·4 + 6510·4) \times 10^4$$

$$= 81·771 \times 10^6 \text{ mm}^4$$

The radius of gyration $k_v = \sqrt{\left(\dfrac{81·771 \times 10^6}{325 \times 10^2}\right)}$

$$= 50·2 \text{ mm}$$

*The slenderness ratio of a strut is defined as the ratio of its length to least radius of gyration*, i.e. slenderness ratio = 7000/50·2 = 139.

SPECIMEN QUESTION 19
State Euler's formula for long struts. Define the symbols used and give the assumptions on which the formula is based. An alloy tube, of 20 mm external diameter and 12 mm bore, 3 m long, extended 1 mm under a steady axial load of 450 kg. By use of Euler's formula, calculate the crippling load for this member used as a pin-jointed strut.

SOLUTION
Euler's formula:

$$P_e = \frac{\pi^2 EI}{L^2}$$

The assumptions made in the derivation of this formula are:

(*a*) the strut is initially perfectly straight;
(*b*) the load is supplied axially;
(*c*) the strut is very long in comparison with its cross-sectional dimensions;
(*d*) the assumptions made in the theory of bending hold good.

Cross-sectional area of tube = $\pi(10^2 - 6^2) = 64\pi$ mm$^2$

$$\text{Stress in tube} = \frac{450 \times 9\cdot8}{64\pi} \text{ N/mm}^2$$

$$\therefore \quad E = \frac{\text{stress}}{\text{strain}} = \frac{450 \times 9\cdot8}{64\pi} \times \frac{3000}{1} = \frac{207\,000}{\pi} \text{ N/mm}^2$$

For *I* of tube about central axis:

the *I* value of a circle about its diameter = $\dfrac{\pi D^4}{64}$ mm$^2$

$$\text{for tube} = \frac{\pi}{64}(20^4 - 12^4) = 2176\pi \text{ mm}^2$$

$$\therefore \text{ Crippling load} = \frac{\pi^2 \times (207\,000/\pi) \times 2176\pi}{3000^2} = 493 \text{ N} \quad (50\cdot3 \text{ kg})$$

The Euler formula is based on the effective length of the strut *L*. This effective length depends in turn on the method of fixing the strut at each end. In specimen question 19 the strut was pin-jointed at both ends, in which case the effective length is the same as the actual length of the strut. Figs. 28(a), (b), (c) and (d) give the relationships between actual length and effective length for different types of end fixing.

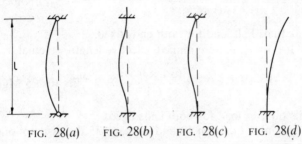

FIG. 28(a)      FIG. 28(b)      FIG. 28(c)      FIG. 28(d)

(a) *Pin jointed at each end:* i.e. ends free to change slope but all other movement prevented. *Effective length L = actual length l*
(b) *Both ends fixed:* i.e. ends firmly clamped so that all movement is prevented. *Effective length L = (actual length l)/2*
(c) *One end fixed, one end pinned. Effective length L = (actual l)/$\sqrt{2} = 0.7\ l$*
(d) *One end fixed, one end completely free. Effective length L = 2 × actual length l*

In practice these figures can be taken only as a guide to the effective length, since it is not normally possible to determine accurately the degree of fixity at the end of a strut.

SPECIMEN QUESTION 20
An alloy rectangular section 24 mm × 50 mm is used as a fixed-ended strut, and is found to carry 1·1 Mg before buckling. Assuming Euler's formula to hold good, what is the length of the strut? $E$ for the alloy used is 180 kN/mm$^2$. What load would the strut carry if (a) one end was pinned, (b) both ends were pinned?

SOLUTION

Least $I$ for strut $= \dfrac{50 \times 24^3}{12} = 57\,600$ mm$^4$

$$P_e = \frac{\pi^2 EI}{L^2}$$

Effective length $L = \pi \sqrt{\left(\dfrac{EI}{P_e}\right)} = \pi \sqrt{\left(\dfrac{180 \times 57\,600}{1\cdot1 \times 9\cdot8}\right)}$

$\qquad\qquad = 3100$ mm $= 3\cdot1$ m

For a fixed-ended strut, effective length = (actual length)/2

$$\therefore \text{ actual length} = \underline{6\cdot2 \text{ m}}$$

(a) If one end were pinned, effective length $= 6\cdot2/\sqrt{2} = 4\cdot38$ m

$$P_e = \frac{\pi^2 \times 180 \times 57\,600}{4380^2} \qquad = 5\cdot4 \text{ kN } (0\cdot55 \text{ Mg})$$

*Note:* equals half load for both ends fixed.

(b) If both ends were pinned, effective length $=$ actual length.

$$P_e = 1\cdot1 \times 9\cdot8 \times \frac{(6\cdot1 \times 0\cdot5)^2}{6\cdot1^2} = 2\cdot7 \text{ kN } (0\cdot275 \text{ Mg})$$

(equals quarter load for both ends fixed).

In specimen question 19 and on page 35 the assumptions of the Euler formula are given. In practice these assumptions rarely hold good and some bending occurs in the strut, since the load may not be concentric or the column may not be perfectly straight, thus causing bending stress. Euler's theory also fails to take yield stress into account.

Several empirical formulae have been developed for practical use.

## Rankine–Gordon formula

If $P$ is the actual load to cause the strut to fail, $P_e$ the load given by Euler's formula to cause buckling, and $P_c$ the force to cause direct compressive failure, then $P$ must be less than either $P_e$ or $P_c$ for a long strut. Also, if $P_e$ is large, then $P_d$ must be small or vice versa. All these requirements are covered by the expression:

$$\frac{1}{P} = \frac{1}{P_e} + \frac{1}{P_c}$$

or

$$\frac{1}{P} = \frac{P_c + P_e}{P_e P_c} = \frac{1 + (P_c/P_e)}{P_c}$$

$$\therefore \qquad P = \frac{P_c}{1 + P_c/P_e}$$

but $P_c = f_c A$    and    $P_e = \dfrac{\pi^2 E I}{L^2}$

$$\therefore \qquad P = \frac{f_c A}{1 + (f_c A L^2/\pi^2 E I)}$$

also $I = A k^2$

$$\therefore \qquad P = \frac{f_c A}{1 + f_c A L^2 / \pi^2 E A k^2}$$

$$= \frac{f_c A}{1 + \frac{f_c}{\pi^2 E}\left(\frac{L}{k}\right)^2}$$

$$= \frac{f_c A}{1 + a\left(\frac{L}{k}\right)^2}$$

where $P$ = actual load to fail strut, $f_c$ = maximum permissible intensity of direct stress, $A$ = area of cross-section, $L$ = effective length of strut, and $k$ = least radius of gyration for cross-section.

$a$ is known as the Rankine constant and depends on the material of the strut. In the equation, $a$ equals $f_c/\pi^2 E$. However, owing to indeterminacies an empirical constant is used. The Rankine constant may be given in tables for pin-ended struts, in which case the effective length of strut $L$ must be used in the formula (depending upon end conditions). Alternatively, the Rankine constant may be given for different end conditions, in which case $L$ is the actual length of strut and $L/k$ is the slenderness ratio.

SPECIMEN QUESTION 21
The symmetrical roof truss shown in Fig. 29(a) is to be constructed from 150 mm × 150 mm timber beams, and loaded as shown. If the yield stress for the timber used is 35 N/mm² and the Rankine constant for pin-ended struts is $\frac{1}{3000}$, check that members X, Y and Z are satisfactory.

SOLUTION
Loading: $\qquad (18\,350/1000) \times 9{\cdot}8 = 180$ kN

First it is necessary to find the forces in members X, Y and Z.
    Reaction at A = reaction at B = $2{\cdot}5 \times 180 = 450$ kN.
    The triangle of forces will be a 30°–60° triangle; therefore force in Y = $450 \times 2 = 900$ kN compression, and force in X = $450\sqrt{3}$ kN tension.

    For the force in Z, consider a truss cut along QQ as shown in Fig. 29(b). Take moments about A:

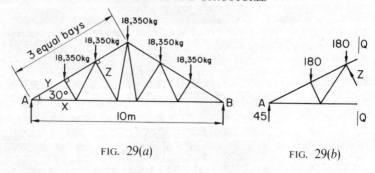

FIG. 29(a)                                    FIG. 29(b)

$$180 \times 5/3 + 180 \times 10/3 = F_Z \times (2/3) \times (10/\sqrt{3})$$
$$F_Z = \frac{(300 + 600)}{20} \times 3\sqrt{3} = 135\sqrt{3} \text{ kN compression.}$$

For member X in tension:

$$f_X = (450\sqrt{3} \times 1000)/(150 \times 150) = 34\cdot6 \text{ N/mm}^2$$

which would be satisfactory with a yield stress of 35 N/mm².

For member Y in compression, from Rankine's formula:

permissible load $P = \dfrac{35 \times 150 \times 150}{1 + \dfrac{1}{3000}\,(L/k)^2}$

$$L = \frac{1000}{3\sqrt{3}} \text{ mm} \quad L^2 = \frac{100\,000\,000}{27} \quad k^2 = \frac{I}{A} = \frac{150 \times 150^3}{12 \times 150 \times 150} = \frac{150^2}{12}$$

$$\therefore \qquad P = \frac{35 \times 150 \times 150}{1 + \dfrac{1}{3000} \times \dfrac{100 \times 10^6}{27} \times \dfrac{12}{150^2}}$$

$$= \frac{35 \times 150 \times 150}{1 + 0\cdot66} = 474\,000 \text{ N} = 474 \text{ kN}$$

Member Y carries 900 kN and would therefore require two members (948 kN permissible load).

For member Z in compression:

$$L = 20\,000/9 \text{ mm}$$
$$1 + \frac{1}{3000} \times \frac{20\,000^2 \times 12}{9^2 \times 150^2} = 1 + 0.88 = 1\cdot88$$

Permissible load B $= 474 \times (1\cdot66/1\cdot88) = 420$ kN

Member Z carries $135\sqrt{3} = 234$ kN, therefore one strut is satisfactory.

SPECIMEN QUESTION 22
Write down the appropriate Euler formula giving the crippling load of a pin-ended strut of uniform cross-section carrying an axial load. Apply this formula to such a strut made from mild steel ($E = 205$ kN/mm$^2$ and yield stress 235 N/mm$^2$) and plot a curve showing the relation between the calculated critical direct stress and the slenderness ratio. Indicate how you would need to modify this curve to give the failing load over the whole range, using Rankine's formula.

SOLUTION
Euler's formula:

$$P_e = \frac{\pi^2 EI}{L^2}$$

$$f_e = \frac{P_e}{A} = \frac{\pi^2 EAk^2}{AL^2}$$

or

$$f_e = \frac{\pi^2 E}{(L/k)^2}$$

If $E = 205$ kN/mm$^2$,

$$f_e = \frac{\pi^2 \times 205\,000}{(L/k)^2} \text{ N/mm}^2$$

From this, with values of $f_e$ from 0 to 300 N/mm$^2$:

| $f_e$(N/mm$^2$): | 0 | 10 | 20 | 30 | 40 | 50 | 60 | 70 | 80 | 90 |
|---|---|---|---|---|---|---|---|---|---|---|
| $L/k$: | ∞ | 450 | 318 | 260 | 225 | 201 | 184 | 170 | 159 | 150 |

| $f_e$(N/mm$^2$): | 100 | 110 | 120 | 130 | 140 | 150 | 160 | 170 | 180 | 190 |
|---|---|---|---|---|---|---|---|---|---|---|
| $L/k$: | 142 | 136 | 130 | 125 | 120 | 116 | 112 | 109 | 106 | 103 |

| $f_e$(N/mm$^2$): | 200 | 210 | 220 | 230 | 240 | 250 | 260 | 270 | 280 | 290 | 300 |
|---|---|---|---|---|---|---|---|---|---|---|---|
| $L/k$: | 101 | 98 | 96 | 94 | 92 | 90 | 88 | 87 | 85 | 84 | 82 |

*Note:* when $L/k = 0$, $f_e = \infty$

These results are shown in graphical form in Fig. 30. The yield stress is 235 N/mm$^2$ and as $L/k$ does not affect this the direct stress failure will be as shown in Fig. 30. The actual stress at which the strut fails must lie beneath this line and therefore assuming:

$$\frac{1}{f} = \frac{1}{f_e} + \frac{1}{f_c}$$

$$f = \frac{f_c}{1 + \dfrac{f_e}{\pi^2 E}(L/k)^2}$$

$$= \frac{\pi^2 E}{\pi^2 E/f_c + (L/k)^2}$$

(*See* the work on the Rankine formula on pages 40–1.)

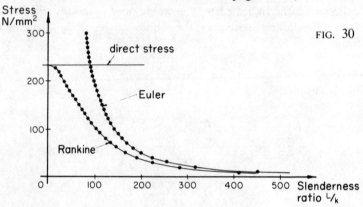

FIG. 30

Compare this with:

$$f_e = \frac{\pi^2 E}{(L/k)^2}$$

It can be seen that the Rankine formula will give a stress lower than the Euler formula for any given value of $L/k$.

The curve for the Rankine formula is also plotted in Fig. 30. Note:

$$\frac{f_c}{E} = \frac{235}{\pi^2 \times 205\,000} = \frac{1}{8600}$$

In practice the value of $a$ for mild steel struts with pinned ends is taken as $\frac{1}{7500}$ and this value has been used to plot the curve.

From the Rankine formula:

$$f = \frac{f_c}{1 + a(L/k)^2}$$

$$\therefore \qquad (L/k)^2 = \left(\frac{f_c}{f} - 1\right)7500$$

$$= \left(\frac{235}{f} - 1\right)7500$$

Note that when $f > f_c$, $(L/k)^2$ is negative and hence has no practical meaning.

With values of $f$ from 0 to 230 N/mm²:

| $f$(N/mm²): | 0 | 10 | 20 | 30 | 40 | 50 | 60 | 70 | 80 |
|---|---|---|---|---|---|---|---|---|---|
| $L/k$: | ∞ | 411·1 | 283·9 | 226·4 | 191·2 | 166·6 | 147·9 | 133 | 120·5 |

| $f$(N/mm²): | 90 | 100 | 110 | 120 | 130 | 140 | 150 | 160 | 170 |
|---|---|---|---|---|---|---|---|---|---|
| $L/k$: | 109·9 | 100·6 | 92·3 | 84·8 | 77·8 | 71·3 | 65·2 | 59·3 | 53·6 |

| $f$(N/mm²): | 180 | 190 | 200 | 210 | 220 | 230 | 235 |
|---|---|---|---|---|---|---|---|
| $L/k$: | 47·9 | 42·1 | 36·2 | 29·9 | 22·6 | 12·8 | 0 |

For the Rankine and Euler formulae to give the same failure stress:

$$\frac{\pi^2 E}{(L/k)^2} = \frac{f_c}{1 + a(L/k)^2}$$

or

$$(L/k)^2 = \frac{\pi^2 E}{f_c - \pi^2 E a}$$

If $a$ is taken to equal $f_c/\pi^2 E$, the curves will cross only when $(L/k^2) = \infty$. If, however, $a$ is taken from tables and $f_c$ is high, $(L/k^2)$ can have a positive value and the curves will cross, e.g. in specimen question 22 take $f_c = 300$ N/mm²:

$$\therefore \ (L/k)^2 = \frac{\pi^2 \times 205\,000}{300 - \left(\dfrac{\pi^2 \times 205\,000}{7500}\right)} = 2\,020\,000/(300 - 270) = 67\,333$$

or   $L/k = 260$ which would give a critical length of strut for both
formulae.

## Perry–Robertson formula
This formula assumes an initial strut curvature of cosine form and is adjusted to take into account imperfections of workmanship and materials as well as eccentricity of load. The formula is:

$$f_c = \frac{f_y + f_e(\eta + 1)}{2} - \tfrac{1}{2}\sqrt{([f_y + f_e(\eta + 1)]^2 - 4 f_y f_e)}$$

where
$f_c$ = compressive stress
$f_y$ = yield stress
$f_e$ = buckling load (Euler)
$\eta$ = imperfection factor

The value of $\eta$ depends upon the slenderness ratio $L/k$ of the strut and may be taken as $0{\cdot}003\,\dfrac{L}{k}$.

In structural steel design $\eta$ is taken as $0{\cdot}3(l/100\,k)^2$ and tables given values of $f_y$ dependant upon $l/k$.

The derivation and use of this formula is beyond the scope of this volume but is given in *Theory of Structures* in this series.

## EXAMINATION QUESTIONS

(In these questions convert all loads to force units, rounded to nearest whole number, before attempting the question.)

1. Fig. 31 shows a system of co-planar loads acting on a flat plate. Calculate the magnitude, sense, direction and position of the equilibrant force.

2. The dimensions and loading of a framework are shown in Fig. 32. The member DE is vertical and the member BD is horizontal. Determine the force in each member and show graphically whether the force is tensile or compressive.

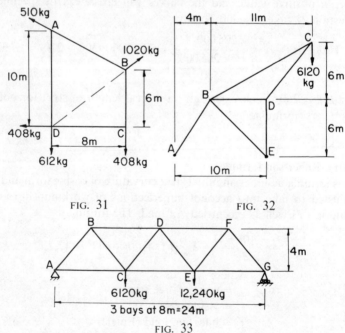

FIG. 31           FIG. 32

FIG. 33

3. Determine the forces in all members of the truss shown in Fig. 33 and state whether they are tensile or compressive.

[Note the similarity to specimen question 14 but here the triangle of forces will be a 45° triangle.]

4. Determine the forces in all members of the truss shown in Fig. 34 and state whether they are tensile or compressive.

[*Note:* the triangle of forces will be a 3–4–5 triangle.]

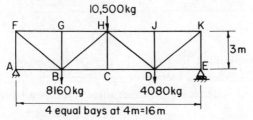

FIG. 34

5. Determine the forces in all members of the truss shown in Fig. 35 and state whether they are tensile or compressive.

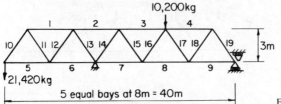

FIG. 35

6. Fig. 36 shows a pin-jointed truss (all joints being indicated by letters) supported at A and B. Determine the force in each of the five members meeting at joint $U_1$ when the truss supports the loads shown in the figure.

(Calculate the force in $U_1L_3$ by the method of sections with a cut vertically between $U_2$ and $U_3$.)

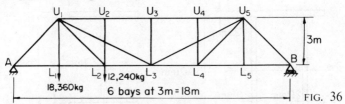

FIG. 36

7. Fig. 37 is an outline diagram of a plane pin-jointed framework supported at A on a pinned support and at B on a roller resting on a plane inclined at 45° to the horizontal. Determine the reactions and the nature and magnitude of the forces in the members X and Z.

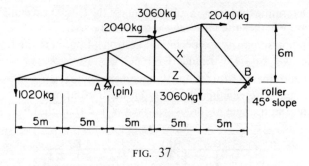

FIG. 37

8. The pin-jointed roof truss in Fig. 38 is subject to the loads shown in the diagram. Determine, graphically, the forces in all the members and state clearly whether these forces are tensile or compressive.

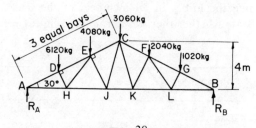

FIG. 38

9. Determine the second moment of area of the section shown in Fig. 39 about a horizontal axis through the centroid.

10. Determine the second moment of area of the section shown in Fig. 40 about a horizontal axis through the centroid.

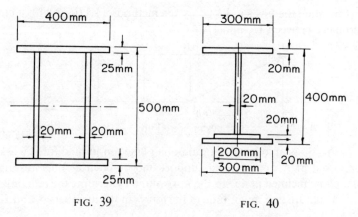

FIG. 39          FIG. 40

11. State the principal assumptions made in the derivation of the Euler formula for the critical load for a slender strut and define the term "ratio of slenderness".

Three steel bars, A, B and C, of circular cross-section, are used as struts as follows.

Bar A: diameter  8 mm,   length   300 mm;   both ends pinned.
Bar B: diameter 16 mm,   length 1200 mm;   both ends fixed.
Bar C: diameter 12 mm,   length  750 mm;   one end pinned and
          the other end fixed.

If the safe load for bar A is taken to be 225 kg, what will be the safe loads for bars B and C, assuming the factor of safety to be the same in each case?

12. The truss shown in Fig. 41 is pinned to a rigid support at B and attached by a mild steel wire from C to A. What diameter would the wire AC have to be.

$f_{steel} = 205 \ \text{N/mm}^2$.

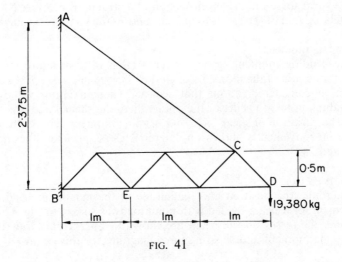

FIG. 41

## CHAPTER 3
# STRUCTURES SUBJECT TO BENDING

IN many structural members the critical stress is due to the bending of the member rather than to direct compression or tension. A common example is the simple beam. Before studying the stress conditions due to bending, it is necessary to analyse a beam under a certain given loading and determine the bending moments. The bending moment at any point in a beam is a function of the shear force at that point, which in turn is a function of the load. Therefore it is necessary to study bending moments and shear force together. The study of stress conditions due to shear will be found in Chapter 4.


## BENDING MOMENT AND SHEAR FORCE DIAGRAMS

### Shear force
The shear force on any given section of a structural member is the algebraic sum of the forces *to one side only* of the section considered.

### Bending moment
The bending moment on any given section of a structure is the algebraic sum of the moments of all the forces *to one side only* of the section considered, about that section. The maximum value of bending moment occurs at the point where the shear force is zero.

The variation of shear force and bending moment across a beam can be conveniently shown on diagrams, as in specimen questions 23 and 24.

SPECIMEN QUESTION 23
Determine the position and magnitude of the maximum bending moment for a beam loaded with point loads as shown in Fig. 42(*a*). (*Note:* the *forces* on the beam are shown in Fig. 42(*b*).) Draw the bending moment and shearing force diagrams for this beam.

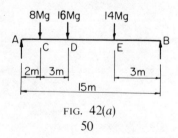

FIG. 42(*a*)

50

SOLUTION

To find the reaction at A, take moments about B:

$$(8 \times 9\cdot8) \times 13 + (16 \times 9\cdot8) \times 10 + (14 \times 9\cdot8) \times 3 = R_A \times 15$$
$$78\cdot4 \times 13 + 156\cdot8 \times 10 + 137\cdot2 \times 3 = 15\,R_A$$
$$R_A = 200 \text{ kN}$$
$$R_B = 78\cdot4 + 156\cdot8 + 137\cdot2 - 200 = 172\cdot4 \text{ kN}$$

Therefore, from the definition, the shear force in the beam between A and C will be 200 kN at all points.

Between C and D the sum of the forces to the left of D will be $200 - 78\cdot4 = 121\cdot6$ kN at all points.

Between D and E the sum of the forces to the left of E will be $200 - 78\cdot4 - 156\cdot8 = -35\cdot2$ kN at all points.

Between E and B the sum of the forces to the left of B will be $200 - 78\cdot4 - 156\cdot8 - 137\cdot2 = -172\cdot4$ kN at all points. These results are shown in Fig. 42(c).

It can be seen that the shear force in the section of beam between E and B can more readily be determined by summing the forces to the right of E, i.e. 172·4 kN, which is the same magnitude as the sum of the forces to the left of E but without the minus sign. Similarly, the sum of the forces to the right of D is $172\cdot4 - 137\cdot2 = 35\cdot2$ kN.

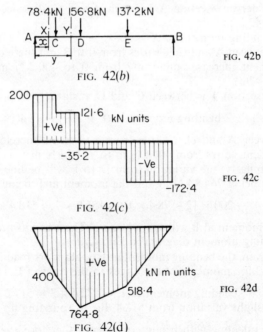

FIG. 42(b)

FIG. 42b

FIG. 42(c)

FIG. 42c

FIG. 42(d)

FIG. 42d

In order to obtain consistent results, therefore, it is necessary to use a sign convention, *which should be observed at all times.* For shear force, if summing the forces to the left of the point, upward forces should be considered positive and downward forces negative. If, however, the forces to the right of the point are summed, then upward forces should be considered negative and downward forces positive.

This will lead to shear force diagrams for simple beams normally having their highest point on the left and descending to the lowest point on the right. Also the diagram is considered positive when above the base and negative when below.

A glance at Fig. 42($c$) shows that the shear force equals zero at point D. Hence the maximum bending moment will be at point D.

∴      Maximum bending moment = bending moment at D

which, from the definition of bending moment, will be:

$$200 \times 5 - 78{\cdot}4 \times 3 = 764{\cdot}8 \text{ kN m}$$

or      maximum bending moment = <u>764·8 kN m at point D.</u>

To construct a bending moment diagram for the whole beam, consider any section $X$ between A and C, distance $x$ from A (*see* Fig. 42($b$)).

Bending moment at $X = 200x$.

Between A and C, $x$ varies from 0 to 2 and therefore the bending moment increases uniformly from 0 to 400 kN m, as shown in Fig. 42($d$).

If section $Y$ is between C and D, distance $y$ from A, then:

$$\text{bending moment at } Y = 200y - 78{\cdot}4(y - 2)$$

Between A and D, $y$ varies from 2 to 5 and therefore the bending moment varies from 400 kN m to 764·8 kN m.

Similarly, the variation from D to E will be linear, the moment at D being 764·8 kN m and the moment at E being:

$$200 \times 12 - 78{\cdot}4 \times 10 - 156{\cdot}8 \times 7 = 518{\cdot}4 \text{ kN m}$$

The moment at B will again be zero. These results are shown on the bending moment diagram in Fig. 42($d$).

Again, the bending moment at E could more readily be calculated by taking moments of the forces to the right of E, i.e.:

$$\text{bending moment at E} = 172{\cdot}4 \times 3 = 517{\cdot}2 \text{ kN m}$$
(slight variation from 518·4 due to rounding up of figures)

When dealing with bending moments, therefore, the result appears the same whether the moments to the left or to the right of the point are summed.

However, a sign convention must be applied to the direction of the moments *and observed at all times*. This convention is that, when summing moments to the left of the point, clockwise moments are positive and anti-clockwise moments negative. When summing moments to the right of the point, clockwise moments are negative and anti-clockwise moments positive.

The bending moment diagram should always be drawn on the "tension face of the beam". A fuller explanation of this is given on page 66. Also, the diagram is considered positive when below the base and negative when above. *It is extremely important to understand the sign convention and apply it at all times, especially for an understanding of more advanced structural analysis.*

## Distributed loads

The loads in specimen question 23 were taken as point loads. In practice it is unlikely that any load would be applied at one small point, but many types of loading approximate to this—column loads, heavy machines, etc.—and can be assumed as point loads. Other types of loading, however, are spread out over a large area and must be treated as distributed load (e.g. concrete floor slab and finishes, stored liquid, granular material, etc.). The method of determining the bending moment and shear force diagrams for a beam subject to distributed load is the same as for point loads, but the final diagrams will appear different.

In the following questions the loads are given in force units.

SPECIMEN QUESTION 24
A beam is loaded with uniformly distributed loads as shown in Fig. 43(*a*). Draw the bending moment and shear force diagrams and calculate the maximum bending moment.

SOLUTION
To determine the reactions, it is necessary to take moments about one of the supports. When dealing with distributed load the total force will be the whole of the load, and its line of action can be taken as acting through the centroid of the distributed load, e.g. take moments about B (*see* Fig. 43(*b*)):

$$18R_A = (30 \times 12) \times 12 + (50 \times 6) \times 3$$
$$R_A = 290 \text{ kN}$$
$$R_B = (30 \times 12) + (50 \times 6) - 290 = 370 \text{ kN}$$

The shear force at A is therefore 290 kN.

At a distance $x$ metres from A between A and C the shear force will be $290 - 30x$. Therefore to find the point where the shear force is zero:

$$290 - 30x_1 = 0$$
$$x_1 = 290/30 = 9\tfrac{2}{3} \text{ m from A}$$

(in this example, $x$ should not be greater than 12 m from A, or the equation no longer applies).

Since the shear force at any point is a function of $x$, the shear force diagram from A to C will be a straight line, crossing the base line $9\tfrac{2}{3}$ m from A (*see* Fig. 43(*c*)).

Also the shear force at B is $-370$ kN and the shear force diagram will be a straight line from C to B.

The shear force at C will be $290 - (30 \times 12) = -70$ kN, or $-370 + (50 \times 6) = -70$ kN.

The bending moment at a point $x$ metres from A, between A and

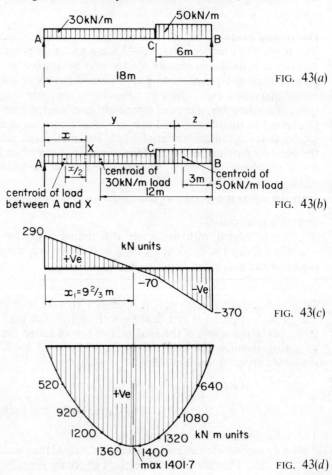

FIG. 43(*a*)

FIG. 43(*b*)

FIG. 43(*c*)

FIG. 43(*d*)

C, will be $290x - (30x \times x/2) = 290x - 30x^2/2$, which will give a parabolic curve.

The maximum bending moment will be at the point of zero shear, $9\frac{2}{3}$ m from A.

$$\text{Maximum bending moment} = 290 \times 9\frac{2}{3} - 30 \times (9\frac{2}{3})^2/2$$
$$= \underline{\underline{1402 \text{ kN m}}}$$

A good exercise at this stage would be to plot the bending moment diagram to scale, using ordinates at 2 m centres.

From A to C, bending moment = $290x - (30x^2)/2$
From C to B, bending moment = $290y - 30 \times 12(y - 6)$
$$- (50(y - 12)^2)/2$$
or from B to C, bending moment = $370z - (50z^2)/2$ which would give the same answer and is easier to calculate.

$$
\begin{aligned}
&2 \text{ m from A B.M.} = 290 \times \;\;2 - (30 \times \;\;2^2)/2 = \;\;520 \text{ kN m}\\
&4 \text{ m from A B.M.} = 290 \times \;\;4 - (30 \times \;\;4^2)/2 = \;\;920 \text{ kN m}\\
&6 \text{ m from A B.M.} = 290 \times \;\;6 - (30 \times \;\;6^2)/2 = 1200 \text{ kN m}\\
&8 \text{ m from A B.M.} = 290 \times \;\;8 - (30 \times \;\;8^2)/2 = 1360 \text{ kN m}\\
&10 \text{ m from A B.M.} = 290 \times 10 - (30 \times 10^2)/2 = 1400 \text{ kN m}\\
&12 \text{ m from A B.M.} = 290 \times 12 - (30 \times 12^2)/2 = 1320 \text{ kN m}\\
&14 \text{ m from A B.M.} = 290 \times 14 - 30 \times 12(14 - 6)\\
&\qquad\qquad\qquad\qquad\quad - (50(14 - 12)^2)/2 = 1080 \text{ kN m}\\
&\text{or} \qquad\qquad\;\; \text{B.M.} = 370 \times \;\;4 - 50 \times 4^2/2 \;\;\;\;= 1080 \text{ kN m}\\
&16 \text{ m from A B.M.} = 370 \times \;\;2 - 50 \times 2^2/2 \;\;\;\;= \;\;640 \text{ kN m}
\end{aligned}
$$

These results are plotted in Fig. 43($d$).

SPECIMEN QUESTION 25
Determine the position and magnitude of the maximum bending moment for the beam shown in Fig. 44($a$), loaded with a triangular load totalling 150 kN.

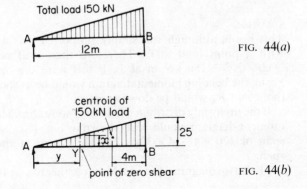

FIG. 44($a$)

FIG. 44($b$)

SOLUTION

The centroid of the load will be 4 m from B. If the intensity of load at B = q,

$$\tfrac{1}{2} \times q \times 12 = 150$$
$$q = 25 \text{ kN/m}$$

or the load varies from 0 at A to 25 kN/m at B (see Fig. 44(b)).

Take moments about B:

$$12R_A = 150 \times 4$$
$$R_A = 50 \text{ kN}$$
$$R_B = 100 \text{ kN}$$

Maximum bending moment occurs at the point of zero shear.
Let the point of zero shear be at point Y, $y$ m from A, and the load intensity at Y equal $x$.

Then downward forces between A and Y must equal upward forces, or $\tfrac{1}{2}xy = 50$.

Also
$$y/x = 12/25$$
∴
$$x = (25/12)y$$
and
$$\tfrac{1}{2} \times (25y/12) \times y = 50$$
$$y^2 = 48$$
$$y = 6 \cdot 93 \text{ m}$$
$$x = (25 \times 6 \cdot 93)12 = 14 \cdot 4 \text{ kN/m}$$

Maximum bending moment = bending moment at Y
$$= 50 \times 6 \cdot 93 - (0 \cdot 5 \times 14 \cdot 4 \times 6 \cdot 93)$$
$$\times (6 \cdot 93/3)$$
$$= \underline{\underline{231 \cdot 2 \text{ kN m}}}$$

SPECIMEN QUESTION 26

Construct the bending moment and shear force diagrams for the beam shown in Fig. 45(a).

SOLUTION

This beam, although still simply supported, has a cantilever, BC, with a point load at C. Therefore there will be a moment of $-60 \times 3 = -180$ kN m at B. If this were the only load on the beam, the bending moment diagram would be as shown in Fig. 45(b) (note that $R_A$ would be downwards).

If the uniformly distributed load only were considered, the bending moment diagram would be as shown in Fig. 45(c), with a maximum span of $120 \times 4 - (4 \times 30) \times 2 = 240$ kN m at the centre of the span.

These two diagrams can be added together to give the final diagram

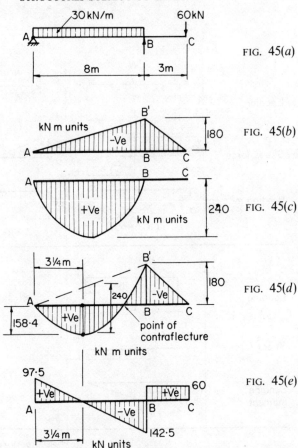

FIG. 45(a)

FIG. 45(b)

FIG. 45(c)

FIG. 45(d)

FIG. 45(e)

for the beam. This is shown in Fig. 45(d), with the parabola displaced to base line AB'. The maximum value of 240 kN m is shown, and is useful for constructing the diagram. The maximum bending moment on the beam (158·4 kN m), however, is calculated as shown at the end of this question.

*Note:* where the bending moment equals zero is known as the *point of contraflecture.*

The shear force diagram can be drawn in the usual manner (Fig. 45(e)), i.e. moments about B:

$$8R_A - (30 \times 8) \times 4 + 60 \times 3 = 0$$
$$R_A = (960 - 180)/8 = 97 \cdot 5 \text{ kN}$$
$$R_B = 240 + 60 - 97 \cdot 5 = 202 \cdot 5 \text{ kN}$$

The point of zero shear will be $97 \cdot 5/30$ m from A $= 3\frac{1}{4}$ m from A.

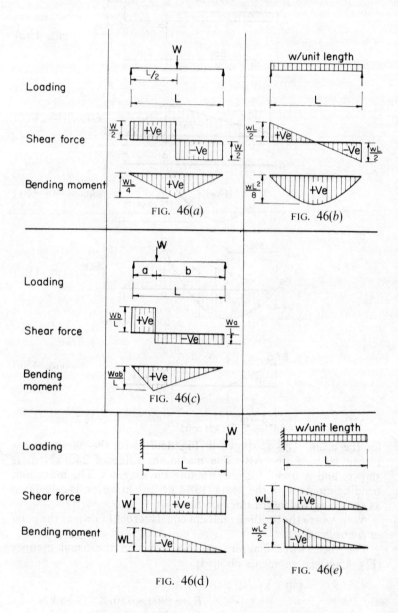

Loading

Shear force

Bending moment

FIG. 46(a)

FIG. 46(b)

Loading

Shear force

Bending moment

FIG. 46(c)

Loading

Shear force

Bending moment

FIG. 46(d)

FIG. 46(e)

$\therefore$     Maximum bending moment $= 97 \cdot 5 \times 3\frac{1}{4} - 30 \times 3\frac{1}{4} \times 3\frac{1}{4}/2$
$$= 158 \cdot 4 \text{ kN m}$$

This is shown in Fig. 45(d).

It would be a useful exercise at this stage to plot the bending moment diagram to scale on a base AB′C, as in Fig. 45(d).

There are a number of standard cases of bending moment and shearing force which are useful to remember. These are shown in Fig. 46: (a) central point load on simple span, (b) uniformly distributed load on simple span, (c) off-centre point load on simple span, (d) point load at end of cantilever, and (e) uniformly distributed load on cantilever.

The importance of being able readily to produce the bending moment and shear force diagram for a structure cannot be over-stressed. The cases dealt with up to this point have been relatively straightforward and, incidentally, very common. The following questions deal with slightly more complex problems, but it should be noted that the same rules are applied throughout.

SPECIMEN QUESTION 27
The beam ABCD shown in Fig. 47(a) is simply supported at B and C. It carries a point load at the free end A, a uniformly distributed load between B and C and an anti-clockwise moment, in the plane of the beam, applied at the free end D. Sketch and dimension the shearing force and bending moment diagrams, and determine the position and magnitude of the maximum bending moment.

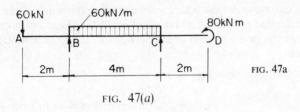

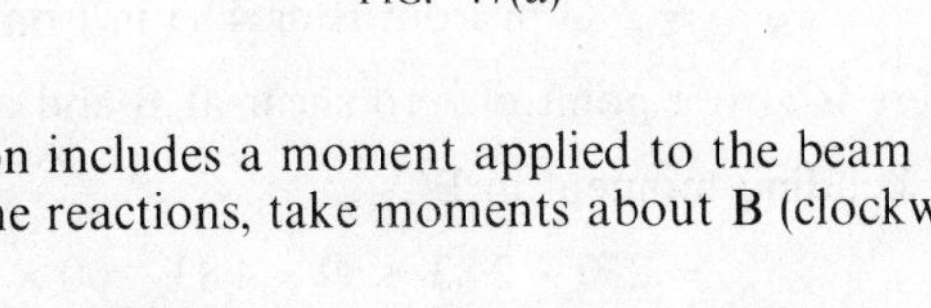

FIG. 47a

FIG. 47(a)

SOLUTION
This question includes a moment applied to the beam at D.

To find the reactions, take moments about B (clockwise about B positive):

$$-60 \times 2 + 60 \times 4 \times (4/2) - 4R_C - 80 = 0$$
$$R_C = 70 \text{ kN}$$

Now take the moments about C (clockwise about C positive):

$$-60 \times 6 + 4R_B - 60 \times 4 \times (4/2) - 80 = 0$$
$$R_B = 230 \text{ kN}$$

*Note:* total vertical load on beam = $60 + 60 \times 4 = 300$ kN
and $$R_B + R_C = 230 + 70 = 300 \text{ kN}$$

i.e. $\Sigma V = 0$ as in previous (and all) cases, which means that $R_B$ could have been found in the usual way without taking moments about C.

The shearing force diagram for specimen question 27 can now readily be produced by taking forces to the left of the point only, and observing the sign convention. This is shown in Fig. 47(c).

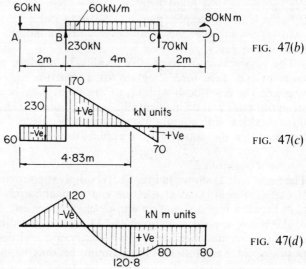

FIG. 47(b)

FIG. 47(c)

FIG. 47(d)

It can be seen that, apart from changing the value of the reactions, the applied moment has no effect upon the shearing force diagram.

Let the point of zero shear between B and C be $x$ metres from B. Then:

$$230 - 60 - 60x = 0$$
$$x = 170/60$$
$$= 2.83 \text{ m from B or } 4.83 \text{ m from A}$$

Note that there is also a point of zero shear at B and at C.

Maximum bending moment in BC

$$= 230 \times 2.83 - 60 \times 4.83 - 60 \times 2.83$$
$$\times (2.83/2)$$
$$= +120.8 \text{ kN m at } 4.83 \text{ m from A.}$$

For the rest of the bending moment diagram:

Bending moment at B = $-60 \times 2 = -120$ kN m
Bending moment at C = $-60 \times 6 + 230 \times 4 - 60 \times 4 \times 2 = +80$ kN m

*Note:* the bending moment at C could more easily be found by taking the moment to the right of C, i.e. +80 kN m.

The bending moment diagram will then be as shown in Fig. 47(*d*). Again, it would be a useful exercise to plot this bending moment diagram to scale.

SPECIMEN QUESTION 28

Figure 48(*a*) shows a beam ABC, 5·5 metres in length, which is supported on a pin joint at A and on rollers at B. At a point D on the beam a vertical arm DE is rigidly connected to it. Sketch diagrams of shearing force and bending moment, indicating maximum values, when the loading shown in the figure is applied.

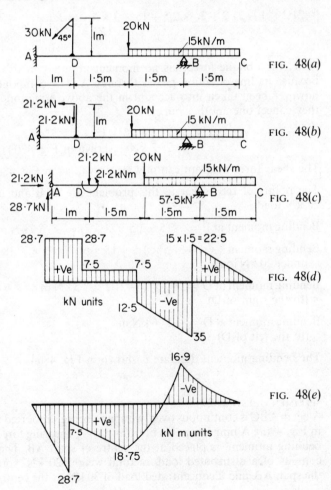

FIG. 48(*a*)

FIG. 48(*b*)

FIG. 48(*c*)

FIG. 48(*d*)

FIG. 48(*e*)

SOLUTION

In this question the oblique force on the vertical arm is a new departure. First split the 30 kN force into horizontal and vertical components, as shown in Fig. 48(b). Now the $30/\sqrt{2} = 21 \cdot 2$ kN vertical component can be treated as an anti-clockwise moment at D of $21 \cdot 2 \times 1 = 21 \cdot 2$ kN m (see Fig. 48(c)). There must also be a horizontal force of 21·2 kN at reaction A (the only place it could be resisted). This horizontal reaction, however, will have no effect upon the vertical shearing force or the bending moment in the beam, because it is at right angles to the vertical and has no lever arm about the beam. Ignoring the horizontal reaction, this is now similar to specimen question 27. Moments about A (clockwise about $A + V_e$):

$$21 \cdot 2 \times 1 - 21 \cdot 2 + 20 \times 2 \cdot 5 + 15 \times 3 \times 4 = 4\ R_B$$
$$R_B = 57 \cdot 5 \text{ kN}$$
(shown in Fig. 48(c))

*Note:* the oblique force has no moment about A in this question because its line of action passes through A. The components have however been taken into account in the above equation, although they cancel one another out.

$$\text{Vertical } R_A = 21 \cdot 2 + 20 + 15 \times 3 - 57 \cdot 5$$
$$= 28 \cdot 7 \text{ kN} \quad \text{(shown in Fig. 48(c))}$$

The shear force diagram can now be drawn, as in Fig. 48(d).

The points of zero shear in this problem are at B and under the 20 kN load.

Bending moment at B $= -15 \times 1 \cdot 5 \times (1 \cdot 5/2) = -16 \cdot 9$ kN m

Bending moment under 20 kN load $= 57 \cdot 5 \times 1 \cdot 5 - 15 \times 3 \times (3/2) = +18 \cdot 75$ kN m

Bending moment at D $= 28 \cdot 7 \times 1 - 21 \cdot 2 = 7 \cdot 5$ kN m
(to the right of D)

Bending moment at D $= +27 \cdot 6$ kN m
(to the left of D)

The bending moment diagram is shown in Fig. 48(e).

SPECIMEN QUESTION 29

A beam ABC is continuous over two spans, being supported as shown in Fig. 49(a). A hinge, capable of transmitting shearing force but not bending moment, is placed at the centre of span AB. The loading consists of a distributed load, of total weight 20 kN, spread over the span AB, and a concentrated load of 30 kN at the centre of span

BC. Sketch the shearing force and bending moment diagrams, indicating on the sketches the magnitudes of all important values.

SOLUTION

Take moments about C:

$$9R_A - 20 \times 6 + 3R_B - 30 \times 1 \cdot 5 = 0$$

or
$$3R_A + R_B = 55 \qquad (1)$$

Moments about B:

$$6R_A - 20 \times 3 + 30 \times 1 \cdot 5 - 3R_C = 0$$
$$2R_A - R_C = 5 \qquad (2)$$

This gives two equations with three unknowns. Taking moments about A would not give a third equation for solution, therefore a third equation must be found in some other way. Now the bending moment at the hinge must equal zero; therefore:

Bending moment at hinge $= 6R_A - 20 \times 1 \cdot 5 = 0$

$\therefore \qquad\qquad R_A = 5$ kN

Substitute in (1):

$$3 \times 5 + R_B = 55$$
$$R_B = 40 \text{ kN}$$

Substitute in (2):

$$2 \times 5 - R_C = 5$$
$$R_C = 5 \text{ kN}$$

This method of determining reactions is used in solving three-pin arch problems (*see* Chapter 6). However, in a beam with a hinge it is easy to see how the reactions are determined.

Consider the beam as shown in Fig. 49(*b*), with beam AD being supported at D on beam DBC.

For beam AD, $R_A = 10/2 = 5$ kN ($R_D = 5$ kN and therefore this will cause a 5 kN point load at D on beam DBC).

For beam DBC, moments about C:

$$3R_B = 5 \times 6 + 10 \times 4 \cdot 5 + 30 \times 1 \cdot 5$$
$$R_B = 40 \text{ kN} \qquad\qquad \left.\begin{array}{c}\\ \\ \\ \end{array}\right\} \text{ shown in Fig. 49(}b\text{)}$$
$$R_C = 5 + 10 + 30 - 40 = 5 \text{ kN}$$

The shear force diagram is shown in Fig. 49(*c*).
Point of zero shear $F = (5 \times 6)/20 = 1 \cdot 5$ m from A. Also, at B and E:

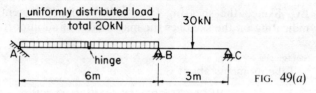

FIG. 49(a)

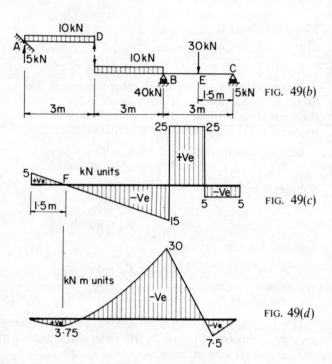

FIG. 49(b)

FIG. 49(c)

FIG. 49(d)

Bending moment at E = 5 × 1·5 = +7·5 kN m
Bending moment at B = 5 × 3 − 30 × 1·5 = −30 kN m
Bending moment at D = 0
Bending moment at point of zero shear *F*
= 5 × 1·5 − (20/6) × 1·5 × (1·5/2)
= +3·75 kN m

The bending moment diagram is shown in Fig. 49(d).

## THEORY OF SIMPLE BENDING

When a beam is subject to load, and therefore bending moments, it will deform. The stresses set up during this deformation must not exceed the permissible bending stresses for the material of the beam.

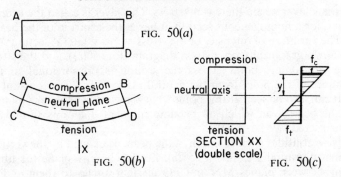

FIG. 50(a)

FIG. 50(b)

SECTION XX
(double scale)

FIG. 50(c)

In order to determine the stresses set up due to bending, it is necessary to examine the deformed beam.

Consider a length of beam AC–BD which is horizontal when unloaded (Fig. 50(a)). When loaded, it deforms as shown in Fig. 50(b).

Assuming that each horizontal layer of material is free to expand or contract individually, it can be seen that the extreme upper layer AB will be shortened and therefore in compression, whilst the extreme lower layer CD will be lengthened and therefore in tension. As the nature of the stress changes from compression to tension between the extreme layers AB and CD, there must be an inter-

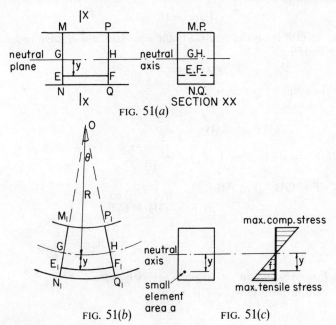

FIG. 51(a)

SECTION XX

FIG. 51(b)

FIG. 51(c)

mediate layer where there is no stress. This layer is called the *neutral plane*; for a homogeneous section, it lies at the centroid of the beam.

The variation of intensity of the stress across a lateral section $XX$ is shown graphically on a stress diagram (Fig. 50(c)), and the stress intensity at any point across the section is proportional to the distance of the point from the neutral plane $y$, or $y/f$ is constant.

It can also be seen at this point what is meant by "the tension face of the beam," on which all bending moment diagrams should be constructed (*see* page 53).

Now consider a short length of the beam as defined by the vertical planes MN and PQ in Fig. 51(a). Let EF be any plane (or fibre) lying between planes MN and PQ at right angles to them and at distance $y$ from the neutral axis. Assuming that plane transverse sections (MN and PQ) remain plane and at right angles to the neutral plane after bending, then planes MN and PQ assume the positions $M_1N_1$ and $P_1Q_1$ as shown in Fig. 51(b). These planes will intersect at a point O. Let the angle between these planes equal $\theta$.

The radius of the neutral plane will be OG = OH = $R$.

Fibre EF is changed in length and becomes $E_1F_1$ and subject to stress $f$ (Fig. 51(c)). The radius of fibre $E_1F_1$ is $R + y$ (or $R - y$ above the neutral plane).

Since the length of an arc equals the radius times the angle subtended at the centre,

$$E_1F_1 = (R + y)\theta$$

Also, as GH is on the neutral plane and does not change in length after bending,

$$GH = R\theta$$

Dividing these equations:

$$\frac{E_1F_1}{GH} = \frac{R + y}{R}$$

Also the strain in fibre $E_1F_1 = \dfrac{E_1F_1 - EF}{EF}$

But EF = GH   (Fig. 51(a)),

$$\therefore \text{ strain in fibre } E_1F_1 = \frac{E_1F_1 - GH}{GH} = \frac{E_1F_1}{GH} - 1$$

$$= \frac{R + y}{R} - 1 = \frac{y}{R}$$

But $E = $ stress/strain, or strain $= $ stress$/E = f/E$

$\therefore$ strain in fibre $E_1F_1 = y/R = f/E$

or                    $\underline{\underline{f/y = E/R}}$

Now consider a small element of cross-section, area $a$, at distance $y$ from the neutral axis (Fig. 51($b$)).

The total force on this element = stress × area = $f \times a$.

The moment of this force on the element, taken about the neutral axis, will be $f \times a \times y$.

The sum of the moments of all the forces acting on all the small elements composing the cross-section will be $\sum fay$.

This external moment must be resisted by the material of the beam, and the maximum value it can reach before the beam fails is equal to the *moment of resistance* of the beam.

Therefore when a member is subject to bending, an internal resistance moment will be set up and

$$\text{internal resistance moment } M_R = \sum fay = \sum \frac{f}{y} ay^2$$

As $f/y$ is a constant,

$$M_R = \frac{f}{y} \sum ay^2$$

but $\sum ay^2$ is the second moment of area of the section about the neutral axis $I_c$.

$\therefore$
$$M_R = \frac{f}{y} I_c$$

or
$$\frac{M_R}{I_c} = \frac{f}{y}$$

but
$$\frac{f}{y} = \frac{E}{R}$$

$\therefore$
$$\boxed{\frac{M_R}{I_c} = \frac{f}{y} = \frac{E}{R}}$$

where $M_R$   is the internal resistance moment of the beam, which, for design purposes, is taken as the applied moment at the section.

$\quad\;\; I_c$   is the second moment of area of the section about the neutral axis,

$\quad\;\; f$   is the stress in any fibre distance $y$ from the neutral axis,

$\quad\;\; E$   is Young's modulus of elasticity for the material of the beam,

$\quad\;\; R$   is the radius of the neutral plane.

It is important that this equation is thoroughly understood, since it forms the basis of all analysis of structural members subject to bending.

### SPECIMEN QUESTION 30

Figs. 52(a) and (b) show details of a loaded beam and its cross-section. Calculate:

(a) the maximum bending stress in the beam;
(b) the moment of resistance of the beam at a maximum permissible bending stress of 165 N/mm²;
(c) the radius of curvature at the point of maximum bending moment.

$$E = 205 \text{ kN/mm}^2$$

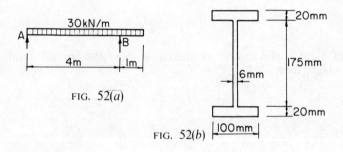

FIG. 52(a)

FIG. 52(b)

### SOLUTION

For loaded beam:

$$R_A = \frac{(5 \times 30) \times 1.5}{4} = 56.25 \text{ kN}$$

Point of zero shear is $56.25/30 = 1.875$ metres from A

∴ maximum span moment $= 56.25 \times 1.875 - 30 \times (1.875^2/2)$
$$= 52.7 \text{ kN m}$$

Moment at B $= -30 \times \frac{1}{2} = 15$ kN m
Maximum moment in beam is $52.7$ kN m

$$I_c \text{ of section} = \frac{100 \times 215^3}{12} - \frac{94 \times 175^3}{12}$$
$$= 40.8 \times 10^6 \text{ mm}^4$$

Depth of neutral axis $= \dfrac{20 + 175 + 20}{2} = 107.5$ mm

$$\frac{M}{I} = \frac{f}{y} = \frac{E}{R}$$

(a) Taking the resistance moment of the beam as equal to the applied bending moment:

$$\frac{52\,700}{40\cdot8 \times 10^6} = \frac{f}{107\cdot5} \qquad \text{(bending moment in kN mm)}$$
$$f = 0\cdot1387 \text{ kN/mm}^2 = \underline{139 \text{ N/mm}^2}$$

This will be the same in tension and compression, as the beam is symmetrical.

(b) If the maximum permissible bending stress is 165 N/mm²,

$$\frac{M}{40\cdot8 \times 10^6} = \frac{165}{107\cdot5}$$
$$M = 62\cdot6 \times 10^6 \text{ N mm} = \underline{62\cdot6 \text{ kN m}}$$

(*Note:* For design this is greater than the applied moment of 52·4 kN m and therefore is safe.)

(c) For the radius of curvature at the point of maximum bending moment:

$$\frac{52\,700}{40\cdot8 \times 10^6} = \frac{205}{R}$$
$$R = \frac{205 \times 40\cdot7 \times 10^6}{52\,700} = 158\,700 \text{ mm or } \underline{159 \text{ m}}$$

SPECIMEN QUESTION 31
Details are given in Figs. 53(a) and (b) of the section and loading on a simply supported girder of 5·5 metres span. Calculate the maximum tensile and compressive stresses which can occur under these conditions of loading.

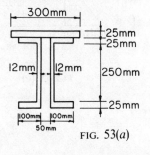

FIG. 53(a)

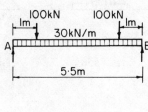

FIG. 53(b)

SOLUTION
For the beam

$$R_A = R_B = \frac{30 \times 5\cdot5 + 100 + 100}{2} = 182\cdot5 \text{ kN}$$

Maximum bending moment is at the centre of the beam

$$M_{max} = 182\cdot5 \times \frac{5\cdot5}{2} - 100 \times \left(\frac{5\cdot5}{2} - 1\right) - 30 \times \frac{5\cdot5}{2} \times \frac{5\cdot5}{4}$$
$$= 502 - 175 - 113\cdot5 = 213\cdot5 \text{ kN m}$$

To find $I_c$: (*see* pp. 35 ff.):

| Portion considered | $b$ ($\times 10^{-1}$) | $d$ ($\times 10^{-1}$) | $A$ ($\times 10^{-2}$) | $\bar{y}$ from top ($\times 10^{-1}$) | $A\bar{y}$ ($\times 10^{-3}$) | $y$ ($\times 10^{-1}$) | $Ay^2$ ($\times 10^{-4}$) | $I_c = bd^3/12$ ($\times 10^{-4}$) |
|---|---|---|---|---|---|---|---|---|
| —— | 30 | 2·5 | 75 | 1·25 | 93·75 | 11·06 | 9174 | 39 |
| – – | 20 | 2·5 | 50 | 3·75 | 187·5 | 8·56 | 3664 | 26 |
| I I | 2·4 | 25 | 60 | 17·5 | 1050 | 5·12 | 1622 | 3125 |
| – – | 20 | 2·5 | 50 | 31·25 | 1562·5 | 18·94 | 17936 | 26 |
| | | | Σ235 | | Σ2893·75 | | Σ32 396 | Σ3216 |

Depth of neutral axis = (2893·75/235) × 10 = 123 mm from top (*see* Fig. 53(*c*)).

$$I_c = (32\,393 + 3216) \times 10^4 = 356 \times 10^6 \text{ mm}^4$$

The maximum tensile stress $f_t$ will occur on the underside of the beam and $y_t$ will be 202 mm (Figs. 53(*c*) and (*d*)).

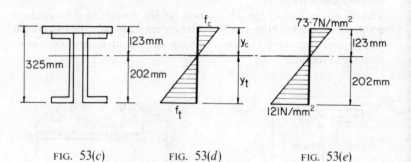

FIG. 53(*c*)        FIG. 53(*d*)        FIG. 53(*e*)

From $\qquad\qquad M/I = f/y$

$$\frac{213 \cdot 5 \times 10^6}{356 \times 10^6} = \frac{f_t}{202} = \frac{f_c}{123}$$

$$f_t = \frac{213 \cdot 5 \times 10^6 \times 202}{356 \times 10^4} = \underline{\underline{121 \ N/mm^2}}$$

$$f_c = 121 \times \frac{123}{202} = \underline{\underline{73.7 \ N/mm^2}}$$

These are shown in the stress diagram, Fig. 53(e).

SPECIMEN QUESTION 32

An unreinforced concrete beam is of the section shown in Fig. 54(a). Determine the maximum moment which could be applied to the section in the plane of the web if the tensile and compressive stresses are limited to 1·4 and 14 N/mm² respectively.

SOLUTION

In this equation the maximum permissible tensile and compressive stresses are not the same. The resistance moment of the beam will be that at which either of these stresses is reached.

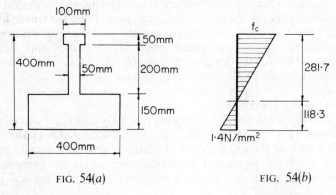

FIG. 54(a)                    FIG. 54(b)

For section:

| Portion considered | $b$ $(\times 10^{-1})$ | $d$ $(\times 10^{-1})$ | $A$ $(\times 10^{-2})$ | $\bar{y}$ from top $(\times 10^{-1})$ | $A\bar{y}$ $(\times 10^{-3})$ | $y$ $(\times 10^{-1})$ | $Ay^2$ $(\times 10^{-4})$ | $I_c = bd^3/12$ $(\times 10^{-4})$ |
|---|---|---|---|---|---|---|---|---|
| ▬ | 10 | 5 | 50 | 2·5 | 125 | 25·67 | 32 950 | 104·2 |
| ❙ | 5 | 20 | 100 | 15 | 1500 | 13·17 | 17 340 | 3333·3 |
| ▬▬ | 40 | 15 | 600 | 32·5 | 19 500 | 4·33 | 11 250 | 11 250 |
| | | | Σ750 | | Σ21 125 | | Σ61 540 | Σ14 687·5 |

Depth of neutral axis = (21 125/750) × 10 = 281·7 mm from top, or 118·3 mm from bottom.

$$I_{NA} = (61\,540 + 14\,687) \times 10^4 = 762 \cdot 27 \times 10^6 \text{ mm}^4$$
$$\frac{M}{I} = \frac{f}{y} \quad \text{or} \quad M = \frac{f\,I}{y}$$

For tension face of beam (bottom):

$$M = \frac{1 \cdot 4 \times 762 \cdot 27 \times 10^6}{118 \cdot 3} = 9 \cdot 02 \text{ kN}$$

For compression face of beam (top):

$$M = \frac{14 \times 762 \cdot 27 \times 10^6}{281 \cdot 7} = 37 \cdot 88 \text{ kN m}$$

Therefore the maximum moment that can be applied to the beam is 9·02 kN m. Above this moment the beam will fail in tension.

The stress in the compression face of the beam at this moment of resistance can be found by reference to the stress diagram (Fig. 54(b)).

$$f_c = (1 \cdot 4/118 \cdot 3) \times 281 \cdot 7 = 3 \cdot 33 \text{ N/mm}^2$$

which is well below the maximum permissible compressive stress.

Since concrete will take high compressive stress but negligible tensile stress, beams in concrete are normally reinforced with steel. The tensile strength of the concrete is then ignored and the tensile forces are resisted by the steel reinforcement. A brief introduction to reinforced concrete analysis is given at the end of this chapter (page 84) but for the design details see *Design of Reinforced Concrete Elements* by R. W. Clements in this series.

SPECIMEN QUESTION 33

A horizontal cantilever 1·25 m long has a T-shaped cross-section, as shown in Fig. 55, and carries a uniformly distributed load along the full length of the top flange. Calculate the greatest intensity of the load which can be carried if the maximum tensile and compressive stresses are not to exceed 30 N/mm² and 90 N/mm² respectively.

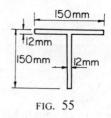

FIG. 55

SOLUTION

Let intensity of loading be $w$ N/m.

Maximum applied moment $= w \times (1.25/2) = 0.781 \, w$ N/m.

| Portion considered | $b$ ($\times 10^{-1}$) | $d$ ($\times 10^{-1}$) | $A$ ($\times 10^{-2}$) | $\bar{y}$ from top ($\times 10^{-1}$) | $A\bar{y}$ ($\times 10^{-2}$) | $y$ ($\times 10^{-1}$) | $Ay^2$ ($\times 10^{-4}$) | $I_c$ ($\times 10^{-4}$) |
|---|---|---|---|---|---|---|---|---|
| — | 15 | 1·2 | 18 | 0·6 | 10·8 | 3·59 | 232 | 2·16 |
| ↑ | 1·2 | 13·8 | 16·56 | 8·1 | 134·1 | 3·91 | 253·7 | 263 |
| | | | $\Sigma 34{\cdot}56$ | | $\Sigma 144{\cdot}9$ | | $\Sigma 485{\cdot}7$ | $\Sigma 265{\cdot}16$ |

Depth of neutral axis $= (144.9/34.56) \times 10 = 41.9$ mm from the top, or 108·1 mm from the bottom.

$$I_{NA} = (485.7 + 265) \times 10^4 = 7.51 \times 10^6 \text{ mm}^4$$

$$\frac{M}{I} = \frac{f}{y} \quad \text{or} \quad f = \frac{My}{I}$$

The top of the cantilever will be in tension

$$f_t = \frac{781w}{7.51 \times 10^6} \times 41.9 = 0.0044w \text{ N/mm}^2$$

The bottom of the cantilever will be in compression

$$f_c = \frac{781w}{7.42 \times 10^6} \times 108.1 = 0.0114w \text{ N/mm}^2$$

If $\qquad f_t = 30 \text{ N/mm}^2$:

$\qquad 0.0044w = 30$

$\qquad\qquad w = 6818 \text{ N/m}, \quad \text{or} \quad 6.82 \text{ kN/m}$

If $\qquad f_c = 90 \text{ N/mm}^2$:

$\qquad 0.0114w = 90$

$\qquad\qquad w = 7895 \text{ N/m}, \quad \text{or} \quad 7.90 \text{ kN/m}$

Therefore the beam would fail in tension and the maximum permissible load is 6·82 kN/m.

Further examples of this type of question are given at the end of this chapter and should now be attempted.


## COMBINED BENDING AND DIRECT STRESS

In many cases a structural member may be subject to both direct stress and bending stress at the same time. Examples are eccentrically loaded columns and pre-stressed beams; also bearing pressure beneath the base of a cantilever retaining wall may be solved in this way.

The total stress in the member can be found by summing the separate stresses, taking care with the sign convention. This can conveniently be shown with stress diagrams as follows.

Consider a member of cross-sectional area $A$ and second moment of area about the neutral axis $I$. Let this member be subject to a direct thrust $F$ acting at eccentricity $e$ and causing moment $My = F \times e$ as shown in Fig. 56($a$). The bending stress in the member (cross-section as in Fig. 56($b$)) will vary from $My_c/I$ compressive at the top to $My_t/I$ tensile at the bottom. This is shown in Fig. 56($c$).

The direct stress in the member $= F/A$ and will be the same for any part of the cross-section. This is shown in Fig. 56($d$).

The combination of these two diagrams gives the final stress diagram shown in Fig. 56($e$).

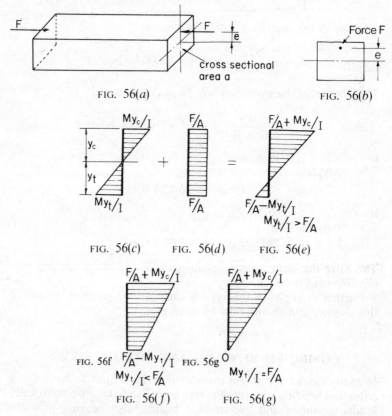

FIG. 56($a$)

cross sectional area a

Force $F$

$e$

FIG. 56($b$)

$My_c/I$    $F/A$    $F/A + My_c/I$

$y_c$

$y_t$

$My_t/I$    $F/A$    $F/A - My_t/I$

$My_t/I > F/A$

FIG. 56($c$)     FIG. 56($d$)     FIG. 56($e$)

$F/A + My_c/I$      $F/A + My_c/I$

FIG. 56f   $F/A - My_t/I$   FIG. 56g   O

$My_t/I < F/A$      $My_t/I = F/A$

FIG. 56($f$)      FIG. 56($g$)

The combined diagram shown in Fig. 56($e$) is for the case where $My_t/I$ is greater than $F/A$.

The combined diagram for $My_t/I$ less than $F/A$ is shown in Fig. 56($f$), and for $My_t/I$ equal to $F/A$ in Fig. 56($g$).

SPECIMEN QUESTION 34

Fig. 57($a$) shows a 600 mm square column which has a bracket carrying a 60 kN load at 200 mm from the face of the column. Calculate the greatest compressive and tensile stresses in the cross-section and draw a diagram to show the stress distribution across the section. The weight of the column should be ignored.

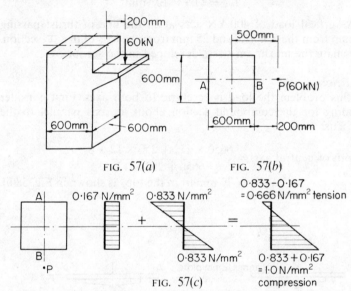

FIG. 57($a$)    FIG. 57($b$)

FIG. 57($c$)

SOLUTION

A cross-section through the column is shown in Fig. 57($b$).

$$I = (600 \times 600^3)/12 = 10 \cdot 8 \times 10^9 \text{ mm}^4$$
$$y = 600/2 = 300 \text{ mm}$$

Owing to the 60 kN load there will be a direct compressive stress on the section $= 60\,000/(600 \times 600) = 0 \cdot 167 \text{ N/mm}^2$.

Owing to the moment of $60 \times 500$ kN mm, the section will be in tension along the face marked A and tensile stress on fibre A

$$= \frac{60\,000 \times 500 \times 300}{10 \cdot 8 \times 10^9} = 0 \cdot 833 \text{ N/mm}^2$$

Face B will be in compression due to bending, the bending stress on fibre B

$$= \frac{60\,000 \times 500 \times 300}{10 \cdot 8 \times 10^9} = 0 \cdot 833 \text{ N/mm}^2$$

The stress diagrams are as shown in Fig. 57(c).

SPECIMEN QUESTION 35

Fig. 58(a) shows the cross-section of a short column, made from a 200 mm × 150 mm ⊥ section, with a 250 mm × 12 mm plate welded to one flange. For 200 mm × 150 mm ⊥ section:

$$area = 6650 \text{ mm}^2$$
$$I_{XX} = 47\cdot6 \times 10^6 \text{ mm}^4$$
$$I_{YY} = 11\cdot9 \times 10^6 \text{ mm}^4$$

A vertical load of 400 kN acts at P, the line of thrust passing 50 mm from the XX axis and 25 mm from the axis of the ⊥ section. Calculate the maximum stress developed in the section.

SOLUTION

In this problem the load is eccentric to both axes. First consider bending for the compound section about an axis parallel to the XX axis.

$$\text{Depth of neutral axis} = \frac{6650 \times 112 + 250 \times 12 \times 6}{6650 + 250 \times 12}$$
$$= 79 \text{ mm from the top, as shown in Fig. 58(b).}$$

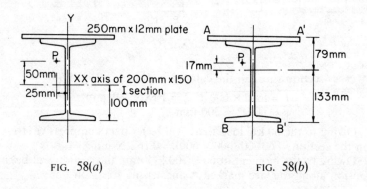

FIG. 58(a)            FIG. 58(b)

$$I = [47\cdot6 \times 10^6 + 6650 \times (112 - 79)^2]$$
$$+ \left[ \frac{250 \times 12^3}{12} + 250 \times 12 \times (79 - 6)^2 \right]$$
$$= 70\cdot89 \times 10^6 \text{ mm}^4$$

400 kN load at P will cause compression at face A A′ and tension at face B B′ due to bending about the neutral axis.

$$\text{At A and A}', f_c = \frac{(400\,000 \times 17) \times 79}{70 \cdot 89 \times 10^6} = 7 \cdot 58 \text{ N/mm}^2$$

$$\text{At B and B}', f_t = \frac{(400\,000 \times 17) \times 133}{70 \cdot 89 \times 10^6} = 12 \cdot 76 \text{ N/mm}^2$$

Next consider bending for the compound section about the YY axis (Fig. 58(c)).

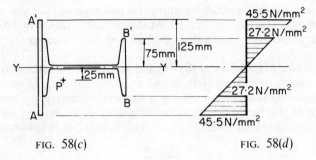

FIG. 58(c)  FIG. 58(d)

$$I_{YY} = 11 \cdot 9 \times 10^6 + \frac{12 \times 250^3}{12} = 27 \cdot 53 \times 10^6 \text{ mm}^4$$

400 kN load at P will cause compression on the AB side of the axis and tension on the A′B′ side.

$$\text{At A}, f_c = \frac{(400\,000 \times 25) \times 125}{27 \cdot 53 \times 10^6} = 45 \cdot 4 \text{ N/mm}^2$$

$$\text{At A}', f_t = \frac{(400\,000 \times 25) \times 125}{27 \cdot 53 \times 10^6} = 45 \cdot 4 \text{ N/mm}^2$$

*Note:*

$$\text{At B}, f_c = \frac{(400\,000 \times 25) \times 75}{27 \cdot 53 \times 10^6} = 27 \cdot 2 \text{ N/mm}^2$$

$$\text{At B}', f_t = \frac{(400\,000 \times 25) \times 75}{27 \cdot 53 \times 10^6} = 27 \cdot 2 \text{ N/mm}^2$$

These are shown on the bending stress diagram (Fig. 58(d)).

Now consider direct compression:

$$f_c = \frac{400\,000}{6650 + 250 \times 12} = 41 \cdot 5 \text{ N/mm}^2$$

which acts at A, A′, B and B′.

Summing up the stress for each point, taking compressive stress positive and tensile stress negative:

$$
\begin{aligned}
\text{At A, } f_c &= & 7\cdot58 + 45\cdot4 + 41\cdot5 &= 94\cdot48 \text{ N/mm}^2 \\
\text{At A', } f_c &= & 7\cdot58 - 45\cdot4 + 41\cdot5 &= 3\cdot68 \text{ N/mm}^2 \\
\text{At B, } f_c &= & -12\cdot76 + 27\cdot2 + 41\cdot5 &= 55\cdot94 \text{ N/mm}^2 \\
\text{At B', } f_c &= & -12\cdot75 - 27\cdot2 + 41\cdot5 &= 1\cdot54 \text{ N/mm}^2
\end{aligned}
$$

The maximum stress developed is $\underline{94\cdot48 \text{ N/mm}^2 \text{ compression at A.}}$

SPECIMEN QUESTION 36

The cross-section through a rectangular pre-stressed concrete beam at mid-span is shown in Fig. 59. At this section a thrust of 270 kN occurs at point A, normal to the cross-section, due to the pre-stressing force.

Calculate (a) the stress at the top and at the bottom surfaces of the beam due to this thrust only; (b) what additional sagging moment can be sustained here if no tension is allowed to occur at the bottom surface of the beam; (c) the compressive stress at the top surface under the combined effect of thrust and moment as in (b) above.

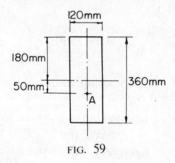

FIG. 59

SOLUTION

(a)          Direct stress $= \dfrac{270\,000}{120 \times 360} = 6\cdot25$ N/mm$^2$

$$
I = \frac{120 \times 360^3}{12} = 36^3 \times 10^4 \text{ mm}^4
$$

Bending stress $f_c = f_t \pm \dfrac{(270\,000 \times 50) \times 180}{36^3 \times 10^4} = \pm 5\cdot21$ N/mm$^2$

∴     stress at top of beam $= 6\cdot25 - 5\cdot21 = \underline{1\cdot04 \text{ N/mm}^2 \text{ compressio}}$

and stress at bottom of beam $= 6\cdot25 + 5\cdot21 = \underline{11\cdot46 \text{ N/mm}^2 \text{ compressi}}$

(b) For just no tension at bottom of beam, additional sagging moment must give tensile stress of 11·46 N/mm² at bottom of beam.

$$\therefore \quad M = \frac{11·46 \times 36^3 \times 10^4}{180} = 29\ 704\ 320 \text{ N/mm}$$
$$= 29·7 \text{ kN m}$$

(c) Sagging moment will give compressive stress of 11·46 N/mm² at the top of the beam, which together with the original pre-stress of 1·04 N/mm² in the top gives:

final compressive stress in the top = 11·46 + 1·04
$$= 12·5 \text{ N/mm}^2$$

This question shows the basic theory used in the design of pre-stressed concrete beams.

SPECIMEN QUESTION 37

Fig. 60(a) is a plan of a three metre length of a mass concrete wall, which has piers at three metre centres. The wall is three metres above ground level and acted on by pressure 1·8 kN/m² which may be considered to act uniformly over the face of the wall as indicated.

Calculate the bearing pressure on the underlying soil (a) at the face AA and (b) at the face BB.

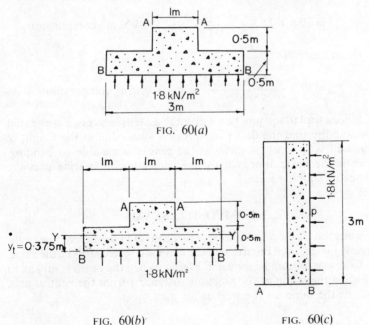

FIG. 60(a)

FIG. 60(b)

FIG. 60(c)

SOLUTION

Fig. 60(b) and (c) shows the section and plan of the wall. The wall will be rotating about axis YY and the loaded face BB will therefore be in tension due to bending.

For section shown in Fig. 60(b)

$$\text{depth of neutral axis} = \frac{1 \times 0.5 \times 0.75 + 3 \times 0.5 \times 0.25}{1 \times 0.5 + 3 \times 0.5} = 0.375 \text{ m from BB}$$

$$I_{NA} = \left[\frac{3 \times 0.5^3}{12} + 3 \times 0.5 \times 0.125^2\right] + \left[\frac{1 \times 0.5^3}{12} + 1 \times 0.5 \times 0.375^2\right]$$
$$= 0.0312 + 0.0234 + 0.0104 + 0.0702 = 0.135 \text{ m}^4$$

Direct compressive stress:

weight of wall $= (3 \times 0.5 + 1 \times 0.5) \times 3 \times 2400 \times 9.8$

$$\therefore \qquad f_c = \frac{(3 \times 0.5 + 1 \times 0.5) \times 3 \times 2400 \times 9.8}{(3 \times 0.5 + 1 \times 0.5)} = 70\,560 \text{ N/m}^2$$
$$= 70.6 \text{ kN/m}^2$$

$$\text{Moment at base} = 3 \times 1.8 \times \frac{3^2}{2} = 24.3 \text{ kN}$$

(a) at face AA

$$f_{AA} = 70.6 + 24.3 \times \frac{(1 - 0.375)}{0.135} = 183 \text{ kN/m}^2 \text{ compression}$$

(b) at face BB

$$f_{BB} = 70.6 - \frac{24.3 \times 0.375}{0.135} = 3 \text{ kN/m}^2 \text{ compression}$$

Since a wall to soil junction will not resist tensile stress, it is essential for stability that the direct compressive stress due to the weight of the wall is at least as great as the tensile stress due to bending. This is the basic approach to the calculation of bearing pressure beneath retaining walls.

## COMPOSITE BEAMS

If a member subject to bending is made up of two different materials, rigidly joined, then the strain at the common surface will be the same.

Consider a beam as shown in Fig. 61. When the beam is subject to bending the strain at the junction, distance $y$ from the neutral axis, will be the same for the steel as for the timber.

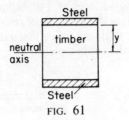

FIG. 61

$$\text{Strain} = \text{stress}/E$$

$$\frac{f_s}{E_s} = \frac{f_t}{E_t}$$

or

$$f_s = f_t \frac{E_s}{E_t}$$

$E_s E_t$ is known as the modular ratio $m$

$\therefore$

$$f_s = mf_t$$

Also at the common surface:

for steel,

$$M_s = \frac{f_s I_s}{y}$$

for timber,

$$M_t = \frac{f_t I_t}{y}$$

$\therefore$ total moment of resistance
$$\begin{aligned}
M &= M_s + M_t \\
&= (f_s I_s + f_t I_t)/y \\
M &= (mf_t I_s + f_t I_t)/y \\
&= \frac{f_t}{y}(mI_s + I_t)
\end{aligned}$$

Therefore, for the compound beam, $mI_s + I_t$ is the "equivalent moment of inertia" of the cross-section, as if it were all made of timber.

An "equivalent timber section" to give this "equivalent moment of inertia" can be taken with all steel dimensions parallel to the neutral axis multiplied by $m$ (hence $I_s$ will automatically be multiplied by $m$).

SPECIMEN QUESTION 38
A timber beam 100 mm wide and 150 mm deep is to be strengthened. Two steel plates 100 mm × 12 mm and 100 mm × 6 mm are adequately secured to it, the thicker plate to the top surface and the thinner to the lower surface, as shown in Fig. 62(a). If the maxi-

mum permissible stress in steel is 140 N/mm² and the value of $E_{steel}/E_{timber} = 20$, calculate the moment of resistance of the strengthened section, assuming that any holes through the steel plates may be neglected, and that the timber will not be overstressed.

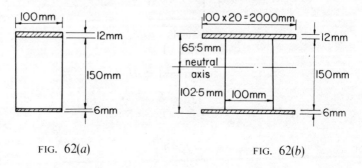

FIG. 62(a)                               FIG. 62(b)

SOLUTION

The equivalent timber section would be as shown in Fig. 62(b).

Depth of neutral axis

$$= \frac{2000 \times 12 \times 6 + 150 \times 100 \times 81 + 2000 \times 6 \times 165}{2000 \times 12 + 150 \times 100 + 2000 \times 6}$$

$$= 65.5 \text{ mm}$$

Equivalent timber $I$

$$= \left[\frac{2000 \times 12^3}{12} + 24\,000 \times 65.5^2\right] + \left[\frac{100 \times 150^3}{12} + 15\,000 \times 21.5^2\right]$$

$$+ \left[\frac{2000 \times 6^3}{12} + 12\,000 \times 99.5^2\right]$$

$$= 288\,000 + 102\,966\,000 + 28\,125\,000 + 6\,933\,750 + 36\,000$$

$$= 257.2 \times 10^6 \text{ mm}^4$$

Equivalent timber stress $= 140/20 = 7 \text{ N/mm}^2$

$$\text{Moment of resistance} = \frac{257.2 \times 10^6 \times 7}{102.5} \text{ N/mm}$$

$$= \underline{17.6 \text{ kN/m}}$$

SPECIMEN QUESTION 39

Fig. 63(a) shows a flitched beam consisting of two timber joists 200 mm × 75 mm and a steel plate 150 mm × 10 mm securely bolted between them. The beam is simply supported on a span of six metres and carries an inclusive uniformly distributed of 900 N/m. Calculate the maximum tensile and compressive stresses in both materials due to this load. $E_{\text{steel}} = 210$ kN/mm$^2$; $E_{\text{timber}} = 8\cdot75$ kN/mm$^2$.

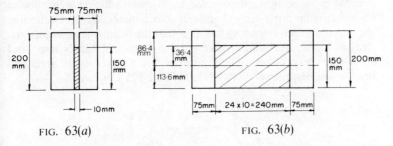

FIG. 63(a)                    FIG. 63(b)

SOLUTION

Maximum applied moment $= (900 \times 6 \times 6)/8 = 4050$ N m

$$m = E_s/E_t = 210/8\cdot75 = 24, \text{ therefore,}$$

the equivalent timber section would be as shown in Fig. 63(b).

$$\text{Height of neutral axis} = \frac{2(75 \times 200 \times 100) + 240 \times 150 \times 125}{(75 \times 200) \times 2 + 240 \times 150}$$

$$= 113\cdot6 \text{ mm}$$

$$\text{Equivalent } I \text{ for timber} = 2\left[\frac{75 \times 200^3}{12} + 75 \times 200 \times 13\cdot8^2\right]$$

$$+ \left[\frac{240 \times 150^3}{12} + 240 \times 150 \times 11\cdot2^2\right]$$

$$= 2[50\,000\,000 + 2\,860\,000]$$
$$+ [67\,500\,000 + 4\,500\,000]$$

$$= 177\cdot72 \times 10^6 \text{ mm}^4$$

For timber:            $f = \dfrac{My}{I}$

(i) In tension, $f_t = \dfrac{4\,050\,000 \times 86\cdot4}{177\cdot72 \times 10^6} = \underline{1\cdot97 \text{ N/mm}^2}$

(ii) In compression, $f_c = \dfrac{4\,050\,000 \times 113\cdot6}{177\cdot72 \times 10^6} = \underline{2\cdot6 \text{ N/mm}^2}$

For steel:

(i) In tension, $f_t = \dfrac{4\,050\,000 \times 86\cdot4}{177\cdot72 \times 10^6} \times 24 = \underline{47\cdot3 \text{ N/mm}^2}$

(ii) In compression, $f_c = \dfrac{4\,050\,000 \times 63\cdot6}{177\cdot72 \times 10^6} \times 24 = \underline{34\cdot8 \text{ N/mm}^2}$

SPECIMEN QUESTION 40

A compound beam (shown in Fig. 64) has a core of 25 mm diameter circular cross-section, surrounded by a smooth-fitting steel tube of hollow circular cross-section and outside diameter 38 mm. Determine the maximum pure bending moment that the beam will sustain, if the stresses in the steel and bronze are not to exceed 125 N/mm$^2$ and 95 N/mm$^2$ respectively. The moduli of direct elasticity for steel and bronze are respectively 210 kN/mm$^2$ and 125 kN/mm$^2$.

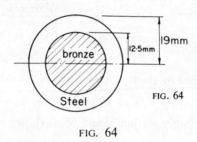

FIG. 64

FIG. 64

SOLUTION

$$\frac{E_s}{E_b} = \frac{210}{125} = 1.68$$

Equivalent $I$ for bronze $= \dfrac{\pi \times 25^4}{64} + 1.68 \times \dfrac{\pi \times (38^4 - 25^4)}{64}$

$$= \frac{\pi}{64}(25^4 + 1.68 \times 38^4 - 1.68 \times 25^4)$$

$$= 0.159 \times 10^6 \text{ mm}^4$$

$$M_{\text{bronze}} = \frac{95 \times 0.159 \times 10^6}{12.5} = 1\,208\,400 \text{ N/mm}$$

$$M_{\text{steel}} = \frac{125/1.68 \times 0.159 \times 10^6}{19} = 622\,650 \text{ N/mm}$$

Maximum bending moment the beam will sustain $= 622\,650$ N/mm or $\underline{\underline{623 \text{ N/m}}}$

## Reinforced concrete

A reinforced concrete beam is a composite beam in which it is assumed that the stress in the concrete is zero.

Fig. 65(a) shows a reinforced concrete beam, breadth $b$, effective depth $d$, area of steel reinforcement $A_s$. Fig. 65(b) shows the strain across the section and Fig. 65(c) the stress across the section.

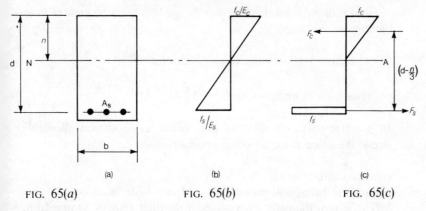

FIG. 65(a)      FIG. 65(b)      FIG. 65(c)

Note that (a) the effective depth of beam is taken to the centre of the reinforcement; (b) the strain is proportional to the distance from the neutral axis; (c) the stress in the steel is uniform.

if $E = \dfrac{\text{stress}}{\text{strain}}$ the values of strain can be marked on the strain diagram (for concrete $f_c/E_c$ for steel $f_s/E_s$).

from the strain diagram

$$\frac{f_c/E_c}{f_s/E_s} = \frac{n}{d-n} \qquad (n = \text{depth to neutral axis})$$

or

$$\frac{f_c}{f_s} = \frac{n}{m(d-n)} \qquad (1)$$

For equilibrium the force resisted by the concrete must equal the force resisted by the steel. From stress diagram:

force resisted by concrete $F_c = \frac{1}{2}f_c n \times b$
force resisted by steel $F_s = f_s A_s$

$$\therefore \qquad \frac{f_c}{f_s} = \frac{2A_s}{n \times b} \qquad (2)$$

from these two equations (1) and (2)

$$\frac{2A_s}{n \times b} = \frac{n}{m(d-n)}$$

which gives an equation from which the depth of the neutral axis $n$ may be found.

Also the resistance moment of the concrete must equal the

resistance moment of the steel (for design this must be greater than the applied moment).

$$\text{resistance moment of concrete} = \tfrac{1}{2}f_c nb \times \left(d - \frac{n}{3}\right)$$

$$\text{resistance moment of steel} = f_s A_s\left(d - \frac{n}{3}\right)$$

In practice only one material may reach its maximum allowable stress, the other material being understressed.

SPECIMEN QUESTION 41

A concrete beam 800 mm deep × 300 mm wide is reinforced with 3 No. 32 mm diameter bars placed with their centres 50 mm from the bottom of the beam as shown in Fig. 66. Find the depth of the neutral axis of the beam.

The maximum permissible compressive stress in the concrete is 30 N/mm² and the maximum permissible tensile stress in the steel is 400 N/mm². Assume full bond develops between the steel and the concrete, that the tensile stress in the concrete is negligible and that the modular ratio $E_s/E_c = 15$.

Calculate the maximum allowable moment of resistance of the section. At this moment of resistance what is the stress in the other material?

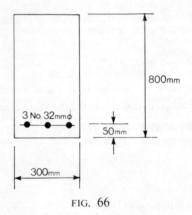

3 No. 32mmⲫ

800mm

50mm

300mm

FIG. 66

SOLUTION

$$A_s = 3 \times \frac{\pi \times 32^2}{4} = 2413 \text{ mm}^2$$

$$\frac{2 \times 2413}{n \times 300} = \frac{n}{15(750 - n)}$$

$$2 \times 2413 \times 15(750 - n) = 300n^2$$
$$n^2 + 241n - 180975 = 0$$
$$n = \underline{\underline{321 \cdot 6 \text{ mm}}}$$

maximum resistance moment of concrete

$$= \tfrac{1}{2} \times 30 \times 321 \cdot 6 \times 300 \times \left(750 - \frac{321 \cdot 6}{3}\right) = 930 \times 10^6 \text{ N mm}$$

maximum resistance moment of steel

$$= 400 \times 2413\left(750 - \frac{321 \cdot 6}{3}\right) = 620 \times 10^6 \text{ N mm}$$

i.e. maximum permissible moment of resistance is when steel reaches maximum stress, i.e. $\underline{\underline{620 \text{ kN m}}}$

at this moment of resistance, stress in concrete

$$f_c = \frac{620 \times 10^6}{\tfrac{1}{2} \times 321 \cdot 6 \times 300\left(750 - \dfrac{321 \cdot 6}{3}\right)} = \underline{\underline{20 \text{ N/mm}^2}}$$

i.e. the concrete is below its maximum permissible stress. For economic design the concrete and steel would both be stressed to their maximum value, but this is unrealistic in practice and it is usual to design for concrete below its maximum.

Although this example shows the basic theory of reinforced concrete design it should be stressed that modern practice uses many design factors and assumptions for which there is no space in this book. For a more detailed coverage see *Design of Reinforced Concrete Elements* in this series.

## EXAMINATION QUESTIONS

1. Fig. 67 (*see* p. 88) shows a simply supported beam carrying the loads indicated. Draw, to a suitable scale, the shear force and bending moment diagrams, marking on all critical values.

2. Calculate the maximum bending moment for the beam shown in Fig. 68 (*see* p. 88) and find the point in the beam where it acts.

3. Draw, to a suitable scale, the shearing force and bending moment diagrams for the beam shown in Fig. 69 (*see* p. 88), and determine the position and magnitude of the maximum bending moment.

4. A beam ABC, 7·5 metres long, is supported at the left hand end, A, and at B, six metres from A. Concentrated loads of 20 kN

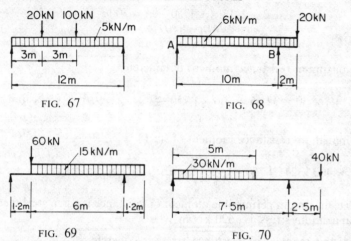

FIG. 67                                    FIG. 68

FIG. 69                                    FIG. 70

and 40 kN act at C and at D, the mid-point of AB, respectively, while a uniformly distributed load of 30 kN per metre acts on the three metre length between D and B.

Sketch the bending moment and shearing force diagrams for the beam, determine the position and magnitude of the point of maximum bending moment, and the point of contraflecture in the span AB.

5. Sketch the bending moment and shearing force diagrams for the beam shown in Fig. 70 and determine the position and magnitude of the maximum bending moment.

6. A beam 7·5 metres long is simply supported at end A and at a support B, six metres from A. If it is loaded as shown in Fig. 71, sketch the shearing force and bending moment diagrams, giving values, including the position and value of the maximum bending moment.

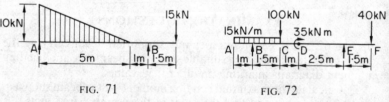

FIG. 71                                    FIG. 72

7. The beam shown in outline in Fig. 72 is simply supported at B and E. Determine the reactions and the values of the bending moment at all important points. Draw, to scale, diagrams of bending moment and shearing force, stating the scales used.

8. Fig. 73 shows the dimensions of, and the loading carried by, a beam ABC. The beam is encastre at A, has a hinge at B and is

supported on a roller bearing at C. Sketch and dimension the shearing force and bending moment diagrams and determine the position and magnitude of the maximum positive and negative bending moments. (*Note*: the fixed end at A may be replaced by a vertical reaction and an unknown moment.)

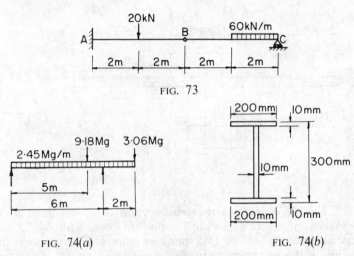

FIG. 73

FIG. 74(*a*)

FIG. 74(*b*)

9. Draw the bending moment and shear force diagrams for the beam shown in Fig. 74(*a*), including important values. If the beam is of constant cross-section over its whole length, as shown in Fig. 74(*b*), calculate the maximum tensile stress that is induced in the material of the beam.

10. A built-up section is composed of two $200 \times 75$ mm channel sections back to back, 25 mm apart, and a $300 \times 12$ mm plate welded across the top to one flange of each channel. Calculate the second moment of area of the section about its neutral axis, and also the moment of resistance about this axis when the maximum permissible stress in the metal is $140$ N/mm$^2$.

For a $200 \times 75$ mm channel section, $I_{XX} = 20 \times 10^6$ mm$^4$; area = $3550$ mm$^2$.

11. The section of a built-up beam is shown in Fig. 75(*a*) (*see* p. 90). The beam has a length of three metres and is constructed by welding two $150$ mm $\times 75$ mm $\times 10$ mm steel angles to a steel plate $460$ mm $\times 12$ mm. The beam is simply supported at its ends and carries a uniformly distributed load on its whole length. Find the magnitude of the maximum permissible load if the maximum stress due to bending is $120$ N/mm$^2$. Neglect the weight of the beam.

For a single angle, the XX axis is shown in Fig. 75(*b*) (*see* p. 90); the cross-sectional area is $2064$ mm$^2$ and $I = 4 \cdot 67 \times 10^6$ mm$^4$.

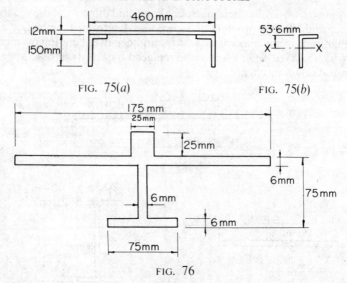

FIG. 75(a)

FIG. 75(b)

FIG. 76

12. Fig. 76 represents the cross-section of an extruded alloy member, which acts as a simply supported beam with the 75 mm wide flange at the bottom. Determine the moment of resistance of the section, if the maximum permissible stresses in tension and compression are respectively 62 and 46 N/mm².

13. The cross-section of a cast iron beam consists of top flange 100 mm × 25 mm, web 300 mm × 25 mm and bottom flange 200 mm × 75 mm. The beam is seven metres long and simply supported at points two metres and 5·5 metres from the left-hand end. Determine the maximum value of uniform vertical load (including self-weight of beam), covering the whole length of beam, consistent with maximum fibre stress due to bending, not exceeding 15 N/mm² in tension or 30 N/mm² in compression. What load would the beam carry if the flanges were reversed?

14. A beam ABC is supported at points A and B and carries a triangular load of 160 kN as shown in Fig. 77(a). If the cross-section of the beam is to be as shown in Fig. 77(b), what will be the maximum tensile and compressive stresses in the material of the beam, and where on the beam will they occur?

15. The arrangement of a ship's davit is shown in Fig. 78(a). The vertical load carried by the davit is 2 Mg. Calculate the greatest compression and tensile stresses due to the load set up at the cross-section AB (Fig. 78(b)). Draw the diagram of stress distribution across AB. (Ignore the self-weight of the davit.)

16. A short aluminium strut consists of two 90 mm × 75 mm × 6 mm angles connected together with a space of 6 mm between them,

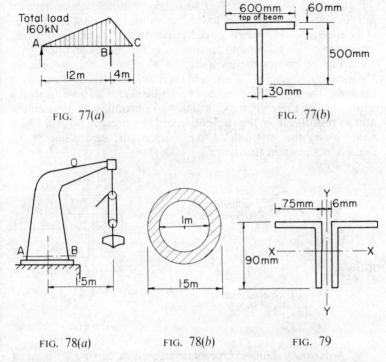

FIG. 77(a)    FIG. 77(b)

FIG. 78(a)    FIG. 78(b)    FIG. 79

as shown in Fig. 79. The load is applied at a point on the YY axis, but not necessarily at the centroid of the section. Determine the limiting positions for the point of application of the load:

(a) for there to be no tensile stress across the section;
(b) for a load of 120 kN and a maximum allowable compressive stress of 75 N/mm². Show the limits on sketches of the section.

17. A short column consists of a 250 mm × 150 mm ⊥ section with a 220 mm × 20 mm plate symmetrically welded to one flange, making the longer dimension of the cross-section 270 mm. A load $W$ is applied to the column, acting through the centroid of the ⊥ section. Calculate the maximum value of $W$ which may be applied to the section.

Maximum permissible stresses: tension = 150 N/mm²; compression = 75 N/mm².

Properties of the ⊥ section: area = 5600 mm²; ⊥ = 65·5 × 10⁶ mm⁴.

18. A rectangular pre-stressed concrete beam is to be used to carry 800 kg/m load across a twelve-metre span. The beam is 600 mm deep by 300 mm wide, and the pre-stress is applied 420 mm from the top of the beam.

$P=59.8$

$f_c = 7.15 N/mm^{-2}$

$f_t = 0.64 N/mm^{-2}$

If the beam is to be stressed to its maximum values in tension and compression, both when loaded and unloaded, what is the value of the pre-stressing force required? What strength must the concrete achieve in compression and tension to satisfy these conditions?

19. Fig. 80 shows the cross-section of a short hollow column. The outside diameter is 200 mm and the diameter of the hollow core is 125 mm. The centre line of the core is displaced 25 mm from the centre line of the column. Calculate the maximum and minimum stresses resulting from the application of a compressive force of 1 MN acting along the longitudinal axis through the centre line of the column as shown in the figure.

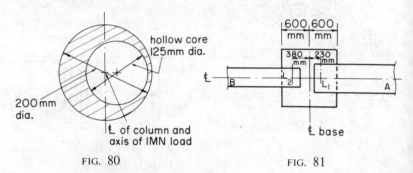

FIG. 80    FIG. 81

20. A masonry dam is trapezoidal in section, having a height of 25 metres, a top width of 2·75 metres and a vertical upstream face. The top water level is 1·2 metres below the crest. If the base width is 14 metres, find the maximum and minimum pressures on the base (a) when the reservoir is empty, (b) when the reservoir is full. The specific gravity of the masonry may be taken as 2·5 and the density of water as 9·8 kN/m³.

21. A masonry pier 1·2 metres square supports two floor girders, A and B (Fig. 81). The loads transmitted by the girders are 400 kN and 200 kN respectively and assumed as concentrated at load points $L_1$ and $L_2$. Determine the stress distribution across the pier and state the maximum compressive stress, ignoring the self-weight of the pier.

Find also the maximum possible addition to the load at $L_1$ so that there may be no tension at any point within the cross-section.

22. A 225 mm × 50 mm steel bar is securely strapped along its length to a 225 mm × 75 mm copper bar. The compound bar is used as a beam 225 mm × 125 mm spanning twelve metres between simple supports and carrying a uniformly distributed load of 1·4 kN/m. Calculate the maximum stresses in the steel and copper (a) when used with the copper and steel side by side, (b) with the steel bar underneath. $E_{steel} = 210$ kN/mm²; $E_{copper} = 125$ kN/mm².

23. A concrete beam 600 mm overall depth and 250 mm wide is reinforced with 2 No. 25 mm bars placed with their centres 40 mm from the underside. The beam is to carry a uniformly distributed load of 12 kN/m over a simply supported span of 6 m. If the modular ratio for steel and concrete is 15, calculate the working stresses in the steel and the concrete.

## CHAPTER 4

# STRUCTURES SUBJECT TO COMPLEX STRESSES

THE only types of stress considered up to this point have been direct stress, buckling stress and bending stress, or certain combinations of these. In this chapter, shear stress and torsional stress are introduced. A structure may be subject to a combination of several different types of stress at the same time, and this leads to some complex stress patterns. Although these may at first appear difficult to analyse, it is often simple to find the maximum stresses, and in an examination paper this type of question is frequently the most straightforward.

In any normal structural analysis, direct force, bending moment and shearing force occur with the greatest frequency, and there is a direct relationship between them.

## RELATIONSHIP BETWEEN LOAD, SHEAR FORCE AND BENDING MOMENT

Consider a short portion of beam AB (length $\delta x$) carrying load $w$/unit length as shown in Fig. 82. (*Note:* the beam is sagging with moments as shown.)

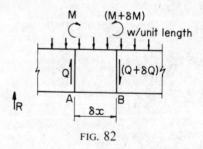

FIG. 82

Let the bending moment at A = M and the bending moment at B = $(M + \delta M)$.

Let the shearing force at A = Q and the shearing force at B = $(Q + \delta Q)$.

For this length to be in equilibrium:

$$Q - w\delta x - (Q + \delta Q) = 0$$

or
$$w = -\frac{\delta Q}{\delta x}$$

Take moments about B. *Note:* when a moment is applied to a beam directly, the total moments to one side only of the point should be added, taking care with the sign convention.

$$M + Q\delta x - w\delta x\frac{\delta x}{2} - (M + \delta M) = 0$$

$w\delta x(\delta x/2)$ is the product of two small quantities and may be ignored.

$$\therefore \qquad\qquad Q\delta x - \delta M = 0$$

or $$Q = \frac{\delta M}{\delta x}$$

The development of this relationship is taken further in Chapter 5, page 146, to include the slope of the beam and its deflection.

## COMPLEMENTARY SHEAR STRESS

Consider a small portion of web of a beam subject to vertical shear stress $q_v$ as shown in Fig. 83($a$). The force acting on each face $= q_v(t \times y)$ giving an anti-clockwise couple acting on the web $= q_v(t \times y) \times x$.

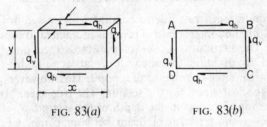

FIG. 83($a$)                FIG. 83($b$)

For equilibrium this must be balanced by a clockwise couple of equal magnitude. This clockwise couple is produced by forces along the horizontal faces, and if $q_h$ is the stress due to these forces,

$$\text{clockwise couple} = q_h(t \times x) \times y$$

since this equals the anti-clockwise couple

$$q_h(t \times x) \times y = q_v \times (t \times y) \times x$$

or $$q_h = q_v$$

i.e. *for equilibrium, a shear stress is automatically accompanied by a shear stress of equal intensity but opposite turning moment in a direction at right angles to that of the original shear stress.*

This method of showing stresses on a diagram and multiplying by the area to give forces is constantly used in calculations involving complex stresses. The convention is normally to show only one face of the material and assume unit thickness, as in Fig. 83(b). The area on which the stress acts is then the length of side × 1 (e.g. $1 \times BC$ or $1 \times AD$). The proof for complementary shear stress can then be abbreviated thus:

$$\text{vertical shearing force} = q_v AD = q_v BC$$
$$\text{anti-clockwise couple} = q_v AD \times AB$$

or
$$\left\{ \begin{array}{l} q_v BC \times AB \\ q_v AD \times DC \\ q_v BC \times DC \end{array} \right\}$$

$$\text{balancing couple} = q_h AB \times AD$$

∴ for equilibrium:

$$q_v AB \times AD = q_v AD \times AB$$
$$\underline{\underline{q_h = q_v}}$$

It is important to be thoroughly conversant with this simple convention when dealing with complex stresses later in this chapter.

## DISTRIBUTION OF SHEAR STRESS

The distribution of stress due to bending is discussed in Chapter 3, together with the construction of bending moment and shearing force diagrams. The distribution of shear force across a section is frequently calculated by dividing the shear force at a point by the cross-sectional area of the member at that point. This, however, gives only the *average* shear stress for the section. The true shear stress is not a constant for all points in the depth of the section.

Consider a short length of beam between planes CE and DF, as shown in Fig. 84(a). The cross section is shown in Fig. 84(b) and is assumed constant. The planes CE and DF are distance $\delta x$ apart and the neutral axis is at depth $y_1$ from the top fibres of the beam.

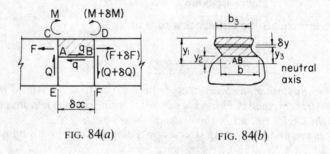

FIG. 84(a)                         FIG. 84(b)

Let the bending moment at section CE = $M$ and the shearing force at this section = $Q$.

Let the bending moment at section DE = $(M + \delta M)$ and the shearing force at this section = $(Q + \delta Q)$.

Consider a layer AB, distance $y_2$ from neutral axis, breadth $b$. Complementary to the vertical shear stress there will be a horizontal shear stress $q$ acting across AB.

Let $F$ be the resultant force on plane AC due to bending moment $M$ and $(F + \delta F)$ the resultant force on plane DB due to bending moment $(M + \delta M)$.

Now consider a thin strip above AB, $\delta y$ thick, distance $y_3$ from the neutral axis and breadth $b_3$.

Let $f$ be the stress due to bending on this strip at section CE:

$$\text{area of cross-section at AC} = \int_{y1}^{y2} b_3 \times \delta y$$

From the theory of bending:

$$\frac{M}{I} = \frac{f}{y} \quad \text{or} \quad f = \frac{My}{I}$$

$$\therefore \qquad \frac{F}{\int_{y1}^{y2} b_3 \delta y} = \frac{My_3}{I}$$

($I$ = second moment of area of the *whole* section about the neutral axis.)

$$F = \frac{M}{I} \int_{y1}^{y2} b_3 \delta y \times y_3$$

but $\int_{y1}^{y2} b_3 \delta y \times y_3$ is the sum of the first moment of area of all the strips above AB about the neutral axis.

$$\therefore \qquad F = \frac{M}{I} a \bar{y} \qquad (1)$$

where $a$ = area of cross-section above AB, and $\bar{y}$ = distance of centroid of area above AB from the neutral axis. Similarly, considering section DB and assuming constant cross-section:

$$(F + \delta F) = \frac{(M + \delta M)}{I} a \bar{y} \qquad (2)$$

Subtracting (1) from (2):

$$\delta F = \frac{\delta M a \bar{y}}{I} \qquad (3)$$

Also, since the element ABDC must be in equilibrium:

$$F + qb\delta x = F + \delta F$$

or

$$\delta F = qb\delta x$$

$$\therefore \qquad qb\delta x = \frac{\delta M\, a\bar{y}}{I}$$

$$q = \frac{\delta M\, a\bar{y}}{\delta x\, Ib}$$

but $\delta M/\delta x = Q$ (*see* page 95),

$$\therefore \qquad q = \frac{Q\, a\bar{y}}{Ib}$$

where   $q$ = horizontal shear stress on plane AB,

         $Q$ = vertical shear force at the section,

         $a$ = area of section above plane AB,

         $y$ = distance from neutral axis to centroid of area above plane AB,

         $I$ = second moment of area of whole section about the neutral axis,

         $b$ = breadth of section at the depth considered.

SPECIMEN QUESTION 42

State the formula which expresses the shearing stress at any depth in the cross-section of a beam; show the symbols used on a diagram. In the case of a beam of rectangular cross-section, show that the maximum shearing stress is 1·5 times the mean shearing stress.

A timber beam of rectangular cross-section is simply supported on a span of four metres and it is required to carry a uniformly distributed load of 7·6 kN/m on the whole span. Determine suitable cross-sectional dimensions for the beam if the maximum allowable values for the stresses in the beam are 7·6 N/mm² in tension and 0·6 N/mm² in shear.

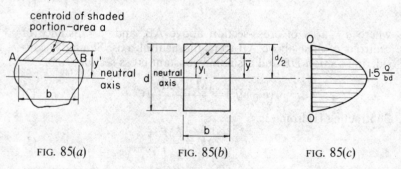

FIG. 85(a)       FIG. 85(b)       FIG. 85(c)

SOLUTION

Shear stress across AB (Fig. 85(a))

$$= q = \frac{Qa\bar{y}}{Ib}$$

$Q$ = vertical shear force,

$I$ = second moment of area of whole section about the neutral axis.

For a rectangular section, depth $d$, breadth $b$, consider the shaded section shown in Fig. 85(b), where AB is $y_1$ above the neutral axis.

Shear stress at AB $= q = \dfrac{Qa\bar{y}}{Ib}$

$$I = \frac{bd^3}{12}$$

$$a = b(d/2 - y_1)$$

$$\bar{y} = \frac{d/2 + y_1}{2}$$

$\therefore$

$$q = \frac{Qb(d/2 - y_1)(d/2 + y_1)/2}{C/12(bd^3/12) \times b}$$

$$= \frac{6Q}{bd^3}\left(\frac{d^2}{2^2} - y_1^2\right)$$

$$= \frac{6Q}{4bd^3}\left(d^2 - 4y_1^2\right) = 1 \cdot 5\frac{Q}{bd^3}\left(d^2 - 4y_1^2\right)$$

which is the equation of a parabola ($\propto y_1^2$) and gives a parabolic curve for shear stress distribution with depth of section as shown in Fig. 85(c)

when                                         $y_1 = \pm d/2$

$$q = 1 \cdot 5\frac{Q}{bd^3}\left(d^2 - \frac{4d^2}{4}\right) = 0$$

maximum value of $q$ is when $y_1 = 0$

when                                         $q = 1 \cdot 5\frac{Q}{bd^3}(d^2 - 0)$

$$= 1 \cdot 5\frac{Q}{bd}$$

The average shear stress across the section is (shear force)/area $= Q/(bd)$, i.e. maximum shearing stress $= 1 \cdot 5 \times$ mean shearing stress.

For the timber beam given:

$$\text{maximum shear force} = \frac{7\cdot6 \times 4}{2} = 15\cdot2 \text{ kN}$$

$$\text{maximum shear stress} = 1\cdot5 \times \frac{15\,200}{bd} = 0\cdot6$$

or $$bd = \frac{1\cdot5 \times 15\,200}{0\cdot6} = 38\,000 \text{ mm}^2$$

Also maximum bending moment $= \dfrac{7\cdot6 \times 4^2}{8} = 15\cdot2 \text{ kN m}$

Second moment of area about the neutral axis $= \dfrac{bd^3}{12} \text{ mm}^4$

Maximum permissible stress in bending $= 7\cdot6 \text{ N/mm}^2$

Depth of neutral axis $= d/2$ mm

$$\frac{M}{I} = \frac{f}{y}$$

$$\frac{15\,200\,000}{bd^3/12} = \frac{7\cdot6}{d/2}$$

$$bd^2 = \frac{15\,200\,000 \times 12}{7\cdot6 \times 2} = 12 \times 10^6 \text{ mm}^3$$

if $$bd = 38\,000 \text{ mm}^2$$

and $$bd^2 = 12\,000\,000 \text{ mm}^3$$

then $$d = \frac{12\,000\,000}{38\,000} = \underline{316 \text{ mm}}$$

and $$b = \frac{38\,000}{316} = \underline{120 \text{ mm}}$$

SPECIMEN QUESTION 43

The section of a 350 mm × 150 mm steel beam may be assumed to consist of two rectangular flanges each 150 mm × 20 mm and a web 10 mm thick. If the beam is subjected to a bending moment of 150 kN m and a shearing force of 250 kN, determine (a) the bending stresses at the inner and outer surfaces of the flanges, (b) the shearing stresses in the web at the junction with the flange and at the neutral axis, (c) the percentage of the total bending moment carried by the flanges, and (d) the percentage of the total shearing force carried by the web (Fig. 86(a), (b) and (c)).

SOLUTION

$$I_{NA} = \frac{150 \times 350^3}{12} - \frac{140 \times 310^3}{12} = 188 \times 10^6 \text{ mm}^4$$

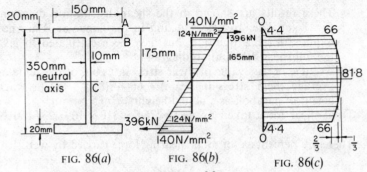

FIG. 86(a)    FIG. 86(b)    FIG. 86(c)

(a) From the theory of bending, $f = \dfrac{My}{I}$

Bending stress at A $= \dfrac{150 \times 10^6 \times 175}{188 \times 10^6} = \underline{140 \text{ N/mm}^2}$

Bending stress at B $= \dfrac{150 \times 10^6 \times 155}{188 \times 10^6} = \underline{124 \text{ N/mm}^2}$

(c) Average stress due to bending in flange $= (140 + 124)/2 = 132$ N/mm$^2$. Force in flange due to bending $= 132 \times 150 \times 20 = 396\,000$ N $= 396$ kN. This can be taken to act at a lever arm of 165 mm from the neutral axis (*see* Fig. 86(b)).

$\therefore$ Moment carried by one flange $= (396 \times 165)/1000 = 65$ kN m and moment carried by two flanges $= 65 \times 2 = 130$ kN m (moment of resistance of flanges).

$\therefore$ percentage of total moment carried by flanges
$$= (130/150) \times 100 = \underline{86.7\%}$$

(b) Horizontal shear stress at A $= 0$.
Horizontal shear stress at B in flange
$$= \frac{250\,000 \times (150 \times 20) \times 165}{188 \times 10^6 \times 150}$$
$$= 4.4 \text{ N/mm}^2$$

Horizontal shear stress at B in web
$$= \frac{250\,000 \times (150 \times 20) \times 165}{188 \times 10^6 \times 10}$$
$$= \underline{66 \text{ N/mm}^2}$$

For horizontal shear stress at the neutral axis:
$a\bar{y} = 150 \times 20 \times 165 + 10 \times 155 \times (155/2) = 615\,125$ mm$^3$

$\therefore$ horizontal shear stress at neutral axis $= \dfrac{250\,000 \times 615\,125}{188 \times 10^6 \times 10}$
$$= \underline{81.8 \text{ N/mm}^2}$$

These results are shown in the shear distribution diagram, Fig. 86(c). The stresses at A and in the flange at B are not required for this question, but have been calculated to give the full shear distribution diagram.

(d) Average of true value of shear stress in web (not the same as average shear stress in beam) $= 66 + \frac{2}{3}(81\cdot8 - 66)$ (average height of parabola is $\frac{2}{3}$ actual height) $= 76\cdot5$ N/mm$^2$

Total shear force taken by web $= 76\cdot5 \times 310 \times 10 = 237\,150$ N $= 237$ kN

Therefore percentage of total shearing force carried by web

$$= \frac{237}{250} \times 100 = \underline{95\%}$$

This question shows clearly that the flange of the beam carries most of the bending moment and the web carries most of the shearing force. In practice, for an $\mathbf{I}$ section, the average shear stress is taken as the shear force divided by the area of the web only, i.e. for specimen question 42,

$$\text{average shear stress in beam} = \frac{250\,000}{350 \times 10} = 71\cdot4 \text{ N/mm}^2$$

SPECIMEN QUESTION 44

Calculate the maximum horizontal shear stress in the beam shown in Fig. 87(a) if it is subjected to a vertical shear force of 120 kN. Sketch the shear stress variation diagram for the section.

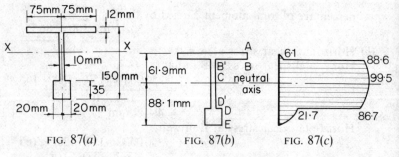

FIG. 87(a)    FIG. 87(b)    FIG. 87(c)

SOLUTION

| Portion of beam | $b$ $\times 10^{-1}$ | $d$ $\times 10^{-1}$ | $A$ $\times 10^{-2}$ | $\bar{y}$ from top $\times 10^{-1}$ | $A\bar{y}$ $\times 10^{-3}$ | $y$ $\times 10^{-1}$ | $Ay^2$ $\times 10^{-4}$ | $I_G$ $\times 10^{-4}$ |
|---|---|---|---|---|---|---|---|---|
| —— | 15 | 1·2 | 18 | 0·6 | 10·8 | 5·56 | 556 | 2·16 |
| – | 1 | 10·3 | 10·3 | 6·35 | 65·4 | 0·19 | 0·37 | 91·06 |
| — | 4 | 3·5 | 14 | 13·25 | 185·5 | 7·09 | 703·75 | 14·29 |
| | | | $\Sigma 42.3$ | | $\Sigma 261\cdot7$ | | $\Sigma 1260\cdot12$ | $\Sigma 107\cdot51$ |

Depth of neutral axis = $(261 \cdot 7/42 \cdot 3) \times 10 = 61 \cdot 9$ mm from top.

$$I_{xx} = (1260 \cdot 4 + 107 \cdot 51) \times 10^4 = 13 \cdot 68 \times 10^6 \text{ mm}^4$$

For points A, B, B', C, D', D, and E, *see* Fig. 87(*b*), $q = \dfrac{Qa\bar{y}}{Ib}$

At A, horizontal shear stress $= 0$

At B, horizontal shear stress $= \dfrac{120\,000 \times (150 \times 12) \times 56 \cdot 1}{13 \cdot 68 \times 10^6 \times 150}$

$\qquad\qquad\qquad\qquad\qquad = 5 \cdot 9 \text{ N/mm}^2$

At B', horizontal shear stress $= \dfrac{120\,000 \times 150 \times 12 \times 56 \cdot 1}{13 \cdot 68 \times 10^6 \times 10}$

$\qquad\qquad\qquad\qquad\qquad = 88 \cdot 6 \text{ N/mm}^2$

At C, $ay = 150 \times 12 \times 56 \cdot 1 + 49 \cdot 9 \times 10 \times \dfrac{49 \cdot 9}{2} = 113\,430 \text{ mm}^3$

$\qquad$ Horizontal shear stress $= \dfrac{120\,000 \times 113\,430}{13 \cdot 68 \times 10^6 \times 10}$

$\qquad\qquad\qquad\qquad\qquad = \underline{99 \cdot 5 \text{ N/mm}^2}$  (maximum)

At D', horizontal shear stress $= \dfrac{120\,000 \times (40 \times 35) \times 70 \cdot 6}{13 \cdot 68 \times 10^6 \times 10}$

$\qquad\qquad\qquad\qquad\qquad = 86 \cdot 7 \text{ N/mm}^2$

*Note:* when below the neutral axis the area below the section is taken.

At D, horizontal shear stress $= \dfrac{120\,000 \times (40 \times 35) \times 70 \cdot 6}{13 \cdot 68 \times 10^6 \times 40}$

$\qquad\qquad\qquad\qquad\qquad = 21 \cdot 7 \text{ N/mm}^2$

At E, horizontal shear stress $= 0$

These are plotted on the stress distribution diagram, Fig. 87(*c*).

## STRESS DUE TO TORSION

Torsional stress is frequently ignored in practical work, but in certain structures—such as a beam supporting a non-continuous cantilever—it can be of considerable importance. Examination questions, however, tend to deal with experimental twisting of circular bars.

The proof of the theory of torsion is similar to the proof of the theory of bending, and it would be a useful exercise for the student to prove it for himself. Only the final result is given here, together with examples of its use.

The expression for torsional stress is:

$$\frac{T}{I_p} = \frac{q}{r} = \frac{G\theta}{l} \quad \left(\text{compare with } \frac{M}{I} = \frac{f}{y} = \frac{E}{R}\right)$$

where $T$ = torque applied to a bar,

    $I_p$ = second moment of area about a polar axis,

    $q$ = shear stress at distance $r$ from the polar axis,

    $G$ = modulus of rigidity, and

    $\theta$ = angle of twist in length $l$ of bar.

This is shown in Fig. 88

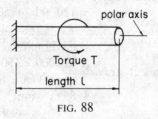

FIG. 88

SPECIMEN QUESTION 45

A mild steel rod of 12 mm diameter is tested in a laboratory. It is first subjected to an axial tension, and under a load of 24 kN its measured extension is 0·15 mm on a 150 mm gauge length.

It is then tested in torsion and, when subjected to a torque of 50 kN m about its longitudinal axis, it rotates through an angle of 0·09 radian on a gauge length of 300 mm.

Calculate the value of (a) Young's modulus of elasticity, (b) shear modulus, (c) Poisson's ratio.

SOLUTION

Much of this question was covered in Chapter 1 (see specimen question 11, page 16). The only new theory is the reference to torsion.

(a) Area of rod $= \dfrac{\pi \times 12^2}{4} = 36\pi$ mm$^2$

Direct stress $f = \dfrac{24\,000}{36\pi} = \dfrac{2000}{3\pi}$ N/mm$^2$

Direct strain $e = \dfrac{0\cdot15}{150} = 0\cdot001$

Young's modulus of elasticity, $E = \dfrac{f}{e} = \dfrac{2000}{3\pi \times 0\cdot001}$

                        $= \underline{\underline{212 \text{ kN/mm}^2}}$

Note: to find the second moment of area about a polar axis, find the second moment of area for the plane section about two central axes at right angles and add them; e.g. for a circle: Fig. 89.

$$I_{XX} = \frac{\pi D^4}{64}$$

$$I_{YY} = \frac{\pi D^4}{64}$$

$$I_p = I_{XX} + I_{YY} = \frac{\pi D^4}{64} + \frac{\pi D^4}{64} = \frac{\pi D^4}{32}$$

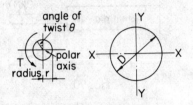

FIG. 89

For specimen question 45:

(b)
$$I_p = \frac{\pi 12^4}{32} = 648\pi \text{ mm}^4$$

$$\frac{T}{I_p} = \frac{G\theta}{l}$$

$$\frac{50\,000}{648\pi} = \frac{G \times 0.09}{300}$$

$$G = \frac{50\,000 \times 300}{648\pi \times 0.09}$$
$$= 82\,000 \text{ N/mm}^2$$
$$= \underline{\underline{82 \text{ kN/mm}^2}}$$

(c)
$$E = 2G(1 + \mu)$$
$$212 = 2 \times 82(1 + \mu)$$
$$\mu = \frac{212}{2 \times 82} - 1 = \underline{\underline{0.29}}$$

SPECIMEN QUESTION 46

A steel shaft, ABC, has a solid circular cross-section of varying diameter. AB is 72 mm diameter and is one metre long; BC is 48 mm diameter and is 0·6 metre long. The end A is securely clamped.

A torque of 2 kN m is applied at C, acting in the opposite direction to an 8 kN m torque at B.

Determine the maximum shearing stress in each part of the shaft and the angle of twist at B and C, relative to A. The modulus of rigidity is 80 kN/mm².

SOLUTION
For section AB: (Fig. 90)

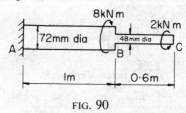

FIG. 90

$$I_p = \frac{\pi \times 72^4}{32} = 840\,000\pi \text{ mm}^4$$

$$\frac{T}{I_p} = \frac{q}{r} = \frac{G\theta}{l}$$

The total torque at A will be $8 - 2 = 6$ kN m

$$\therefore \qquad \frac{6000}{840 \times 10^3 \pi} = \frac{q}{72/2}$$

In AB, maximum $q = \dfrac{6000 \times 36}{840 \times 10^3 \pi} \times 1000 = \underline{\underline{81 \cdot 8 \text{ N/mm}^2}}$

Also

$$\frac{81 \cdot 8}{72/2} = \frac{80\,000 \times \theta_{BA}}{1000}$$

$$\theta_{BA} = \frac{81 \cdot 8 \times 1000}{80\,000 \times 36} = 0 \cdot 0284 \text{ radian}$$

$$= 0 \cdot 0284 \times \frac{180}{\pi} = \underline{\underline{1 \cdot 63°}}$$

For section BC,

$$I_p = \frac{\pi \times 48^4}{32} = 166\pi \times 10^3 \text{ mm}^4$$

The total torque at B (in section BC) will be 2 kN m

$$\therefore \qquad \frac{2000}{166 \times 10^3 \pi} = \frac{q}{48/2}$$

In BC, maximum $q = \dfrac{2000 \times 24}{166 \times 10^3 \pi} \times 1000 = \underline{\underline{92 \text{ N/mm}^2}}$

also $\dfrac{92}{48/2} = \dfrac{80\,000 \times \theta_{CB}}{600}$

$$\theta_{CB} = \dfrac{92 \times 600}{80\,000 \times 24} = 0{\cdot}0288 \text{ radian}$$

$\therefore \qquad \theta_{CA} = \theta_{CB} - \theta_{BA} = 0{\cdot}0288 - 0{\cdot}0284 = 0{\cdot}0004 \text{ radians}$

i.e. C had negligible twist relative to A.

SPECIMEN QUESTION 47

The cross-section of a main shaft at a bearing consists of a central steel core 72 mm diameter, with a bronze sleeve 12 mm thick shrunk on to the core so that there is no slipping between steel and bronze. If at the bearing the torque is 10 kN m, determined the maximum shear stress in steel and bronze due to this torque. For steel, modulus of rigidity = 80 kN/mm². For bronze, modulus of rigidity = 48 kN/mm².

SOLUTION

As for a compound section in bending (*see* Fig. 91), the equivalent polar second moments of area can be found by multiplying by the ratio of the moduli of rigidity, i.e.:

$$\frac{G_s}{G_b} = \frac{80}{48} = 1{\cdot}67$$

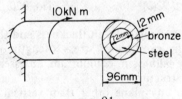

FIG. 91

If all the section were bronze:

$$\text{equivalent } I_p = 1{\cdot}67 \times \frac{\pi \times 72^4}{32} + \frac{\pi(96^4 - 72^4)}{32}$$

$$= \frac{\pi 96^4}{32} + 0{\cdot}67 \times \frac{\pi 72^4}{32} = (2{\cdot}64 + 0{\cdot}56) \times 10^6 \pi$$

$$= 3{\cdot}2\pi \times 10^6 \text{ mm}^4$$

$$\therefore \quad \frac{10\,000\,000}{3\cdot2\pi \times 10^6} = \frac{q_b}{96/2}$$

$$\therefore \text{ Maximum shear stress in brass} = \frac{10 \times 48}{32\pi} = 4\cdot78 \text{ N/mm}^2$$

$$\text{Maximum shear stress in steel} = \frac{10}{32\pi} \times 36 \times 1\cdot67 = 6 \text{ N/mm}^2$$

## COMBINED STRESSES

If a point in a structure is subject to both direct stresses and to shear stress simultaneously, it can be shown on a diagram in the same manner as the convention given in Fig. 83(b), i.e. for a part of a structure ABCD subject to direct compressive stress $f_x$ acting horizontally, direct tensile stress $f_y$ acting vertically and a positive vertical shear stress $q$, the diagram would be as shown in Fig. 92(a).

However, as shown earlier (page 95) there will automatically be a complementary horizontal shear stress of magnitude $q$. This is included in Fig. 92(b).

FIG. 92(a)          FIG. 92(b)

The material is assumed of unit thickness and therefore the *force* on any face is the product of stress × length of side ($f_x \times$ AD, $f_y \times$ DC, etc.). Fig. 92(b) is in equilibrium, an essential requirement for any part of a structure.

If the stress on any plane other than vertical or horizontal is required, this can readily be found. Assume it is required to find the stresses on a plane at angle $\theta°$ to the horizontal as shown in Fig. 92(c).

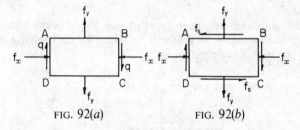

FIG. 92(c)          FIG. 92(d)          FIG. 92(e)

Consider the triangular portion BCE only (Fig. 92($d$)). For this portion to be in equilibrium, there must be a force acting on EB.

Whatever this force may be, it can be resolved into components along the plane and at right angles to it. Let the stresses due to these component forces be $q_1$ and $f_1$ respectively, and assumed acting in the direction shown in Fig. 92($d$).

This portion EBC is now in equilibrium. Resolve *forces* in a direction at right angles to EB:

$$f_1 \times \text{EB} - f_y \times \text{EC} \cos \theta + f_x \times \text{BC} \sin \theta - q \times \text{EC} \sin \theta$$
$$- q \times \text{BC} \cos \theta = 0$$

Divide through by EB:

$$f_1 = f_y \frac{\text{EC}}{\text{EB}} \cos \theta - f_x \frac{\text{BC}}{\text{EB}} \sin \theta + q \frac{\text{EC}}{\text{EB}} \sin \theta + q \frac{\text{BC}}{\text{EB}} \cos \theta$$

But    $\dfrac{\text{EC}}{\text{EB}} = \cos \theta$    and    $\dfrac{\text{BC}}{\text{EB}} = \sin \theta$

$\therefore$    $f_1 = f_y \cos^2 \theta - f_x \sin^2 \theta + 2q \cos \theta \sin \theta$

Resolve *forces* in a direction parallel to plane EB:

$$q_1 \times \text{EB} - f_y \times \text{EC} \sin \theta - f_x \times \text{BC} \cos \theta + q \times \text{EC} \cos \theta$$
$$- q \times \text{BC} \sin \theta = 0$$

Divide by EB    (EC/EB = $\cos \theta$    BC/EB = $\sin \theta$):

$$q_1 = f_y \cos \theta \sin \theta + f_x \sin \theta \cos \theta - q \cos^2 \theta + q \sin^2 \theta$$

or        $q_1 = (f_y + f_x) \sin \theta \cos \theta - q(\cos^2 \theta - \sin^2 \theta)$

but                    $\sin \theta \cos \theta = \frac{1}{2} \sin 2\theta$

and                    $\cos^2 \theta - \sin^2 \theta = \cos 2\theta$

These results may be expressed:

$$f_1 = f_y \cos^2 \theta - f_x \sin^2 \theta + q \sin 2\theta$$
$$q_1 = (f_y + f_x) \frac{\sin 2\theta}{2} - q \cos 2\theta$$

Since $f_y$, $f_x$ and $\theta$ are known, $f_1$ and $q_1$ may be calculated. $f_1$ will be the direct stress on plane EB and $q_1$ the shear stress on plane EB.

It is not of any value to remember these two equations for $f_1$ and $q_1$, but the student should be thoroughly familiar with the manner in which they are derived. No attempt at a sign convention has been made, the directions of forces being arrived at simply by inspection of Fig. 92($d$).

There are, however, certain special cases of stresses on inclined planes which are worth noting, and the above equations have been used to demonstrate these.

*Case 1*

When $f_y = f_x = 0$   (i.e. shear stress $q$ only acts) and $\theta = 45°$

$$f_1 = 0 - 0 + q \sin (2 \times 45°)$$

$\quad$ (sin 90° = 1)

$\therefore \qquad f_1 = q$   ($f_1$ tensile, as shown in Fig. 92($d$))

$$q_1 = 0 - q \cos (2 \times 45°)$$

$\quad$ (cos 90° = 0)

$\therefore \qquad q_1 = 0$

Similarly, it may be shown that if a plane at right angles to EB is taken (Fig. 92($e$)):

on plane FC

$\qquad f_2 = q$   ($f_2$ compressive, as shown in Fig. 92($e$))

and $\qquad q_2 = 0$

i.e. *two complementary shear stresses, on planes perpendicular to each other, are equivalent to direct tensile and compressive stresses of intensity equal to that of the shear stress, on planes inclined at 45° to the shear stress.*

This is the basis of the reason for placing shear reinforcement bars at 45° close to a support in a reinforced concrete beam. These bars take the tensile stress, as concrete has very low tensile strength. The compressive stress at 90° to the bars can be resisted by the concrete.

*Case 2*

When $\qquad\qquad q = 0$   (direct stresses only)

$$q_1 = (f_y + f_x)\frac{\sin 2\theta}{2} - 0$$

$$= (f_y + y_x)\frac{\sin 2\theta}{2}$$

Maximum value of $q_1$ ($q_{max}$) occurs when $\sin 2\theta = 1$

i.e. $\qquad\qquad \theta = 45°$

and $\qquad\qquad q_{max} = \frac{f_y + f_x}{2}$

(as $f_y$ is tensile and $f_x$ is compressive in this example, $f_y + f_x$ is in fact the difference between the vertical and horizontal direct stresses.)

*When direct stresses only act on two planes perpendicular to each other, the maximum shear stresses act on planes at 45° to these planes and are of magnitude equal to half the difference of the direct stresses.*

*Case 3*

When $\qquad q_1 = 0$   (no shear stress on plane EB)

$$q_1 = 0 = (f_y + f_x)\frac{\sin 2\theta}{2} - q \cos 2\theta$$

$$\therefore \qquad q = +(f_y + f_x)\frac{\tan 2\theta}{2}$$

or $\qquad \tan 2\theta = +2q/(f_y + f_x)$

This is an expression in terms of the original stresses from which two values of $\theta$ can be found, differing by $90°$, i.e. two planes at right angles to each other on each of which there is no shear stress. In fact there is also a third plane, mutually at right angles to these two, on which there is no shear stress. A plane across which no shear stress acts is known as a *principal plane*.

The direct stress acting on this plane can be found from the equation:

$$f_1 = f_y \cos^2 \theta - f_y \sin^2 \theta + q \sin 2\theta$$

This is known as the *principal stress* $(f_p)$.

An understanding of principal stress is most important both for advanced practical work and for elementary examinations. Many students find this subject difficult to understand, but if care is taken to grasp thoroughly:

(a) the conventional method of showing stresses on a plane section of unit width,
(b) the conversion of stress to force by multiplying by the length of side,
(c) the resolution of these forces in two directions at right angles,
(d) that all parts of a system must always be in equilibrium,

then problems in combined stress should cause no difficulty. The student should therefore make sure he completely understands the principles of the calculations in the last few pages. Most of the equations derived from these calculations are of little importance and no attempt should be made to remember them.

## PRINCIPAL PLANES AND PRINCIPAL STRESSES

No matter how complex the stresses at any point in a structure may be, there will always be three planes, mutually perpendicular to each other, on which there is no shear stress (i.e. direct stress at right angles to the plane only is acting). These planes are known as *principal planes* and the direct stress acting on each is known as the *principal stress*.

In practice the principal stresses are often due to applied pressures, such as a bar in direct tension, a concrete test cube or a soil sample in a tri-axial apparatus. In these cases the principal stresses are known and the shear stress on some other plane may be required. However, when a shear stress is induced in the material, as in the web of a beam or a bar in torsion, then the principal planes may be required, together with the principal stresses acting on them. For simplicity, the stress on the third plane is usually taken to be zero, thus giving only a two-dimensional problem.

Summing the various stresses which may be given, or required to find at a point in a structure, we have:

(a) direct stresses in two directions at right angles ($f_x$ and $f_y$), together with shear stress ($q$) on the same planes;

(b) maximum and minimum principal stresses ($f_{p1}$ and $f_{p2}$) in two directions at right angles, which, by definition, means no shear stress on these planes;

(c) maximum shear stress ($q_{max}$), which acts on planes at 45° to the principal planes, and is equal to half the difference between the principal stresses $(f_{p1} - f_{p2})/2$. There will also be a direct stress on this plane;

(d) direct and shear stress ($f_1$ and $q_1$) on a plane at any given angle to one of the principal planes.

SPECIMEN QUESTION 48

Explain the terms "principal plane" and "principal stress."

If at a point in a material there exist on planes at right angles compressive stresses $f_x$ and $f_y$, together with complementary shearing stresses $q$ acting across each plane, deduce from first principles an expression giving the magnitudes of the two principal stresses at that point.

Find the values of the shearing stress $q$ and also the minimum principal stress in a case where $f_x = 90$ N/mm², $f_y = 45$ N/mm² and maximum principal stress $= 120$ N/mm², all compressive.

SOLUTION

A "principal plane" is a plane within a material on which there is no shear stress, i.e. all stress is direct stress acting at right angles to the plane. There are always three such planes in a material subject to stress, and these three planes are at right angles to each other.

A "principal stress" is the direct stress acting on the principal plane (see definitions given above).

The magnitudes of the two principal stresses (third principal stress

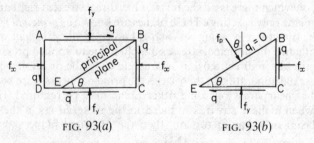

FIG. 93(a)          FIG. 93(b)

equals zero) could be found by substituting in the equations for $\theta$ and $f_p$ developed earlier. However, it is easier to work from first principles and the question requires that this should be done.

Fig. 93(*a*) shows the stresses acting at the point in the material.

If one principal plane makes and angle $\theta$ with the horizontal, the stresses acting on EBC are shown in Fig. 93(*b*).

Resolve forces horizontally:

$$f_p \, \text{EB} \sin \theta = f_x \, \text{BC} + q \, \text{EC}$$

Divide by EB (BC/EB = $\sin \theta$ EC/EB = $\cos \theta$):

$$f_p \sin \theta = f_x \sin \theta + q \cos \theta \qquad (1)$$

or

$$q = (f_p - f_x) \tan \theta \qquad (2)$$

resolve forces vertically:

$$f_p \, \text{EB} \cos \theta = f_y \, \text{EC} + q \, \text{BC}$$

Divide by EB:

$$f_p \cos \theta = f_y \cos \theta + q \sin \theta \qquad (3)$$

$$q = (f_p - f_y) \cot \theta \qquad (4)$$

Divide equation (1) by equation (3):

$$\frac{f_p \sin \theta}{f_p \cos \theta} = \frac{f_x \sin \theta + q \cos \theta}{f_y \cos \theta + q \sin \theta}$$

$$f_x \sin \theta \cos \theta + q \cos^2 \theta = f_y \cos \theta \sin \theta + q \sin^2 \theta$$

$$(f_y - f_x) \cos \theta \sin \theta = q(\cos^2 \theta - \sin^2 \theta)$$

but

$$\cos \theta \sin \theta = \sin 2\theta / 2$$

and

$$\cos^2 \theta - \sin^2 \theta = \cos 2\theta$$

$\therefore$

$$(f_y - f_x)\frac{\sin 2\theta}{2} = q \cos 2\theta$$

or

$$\underline{\underline{\tan 2\theta = \frac{2q}{f_y - f_x}}}$$

This will give two values differing by 90° and these will be the two principal planes. (This is the same as the expression developed on page 111, the difference in signs being due to the difference in the direction of stresses.)

Multiply equation (2) by equation (4):

$$q^2 = (f_p - f_x) \tan \theta \, (f_p - f_y) \cot \theta$$

$$q^2 = f_p^2 - f_p f_y - f_p f_x + f_x f_y$$

or

$$f_p^2 - (f_p + f_x)f_p + (f_x f_y - q^2) = 0 \qquad (5)$$

This is a quadratic in $f_p$ which can be solved in the usual way, i.e.

$$f_p = \frac{(f_y + f_x) \pm \sqrt{(f_y + f_x)^2 - 4(f_x f_y - q^2)}}{2}$$

$$= \tfrac{1}{2}[(f_y + f_x) \pm \sqrt{f_y^2 + f_x^2 + 2f_x f_y - 4f_x f_y + 4q^2}]$$

$$= \tfrac{1}{2}[(f_y + f_x) \pm \sqrt{(f_y - f_x)^2 + 4q^2}] \qquad (6)$$

This will give two values of principal stress, the maximum value being when the $+$ sign is used and the minimum when the $-$ sign is used. *Note:* The maximum shear stress on the material will be on a plane at $45°$ to the principal planes and its magnitude will be half the difference between the principal stresses (*see* page 110), i.e.:

maximum shear stress,

$$q_{max} = \frac{\frac{1}{2}[(f_y + f_x) + \sqrt{(f_y - f_x)^2 + 4q^2}] - \frac{1}{2}[(f_y + f_x) - \sqrt{(f_y - f_x)^2 + 4q^2}]}{2}$$
$$= \frac{1}{2}\sqrt{(f_y - f_x)^2 + 4q^2} \quad \text{acting at } (\theta \pm 45°) \text{ to the horizontal.}$$

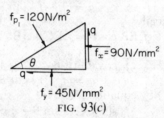

FIG. 93(c)

The last part of specimen question 48 can be solved by substituting in equation (5) (*see* Fig. 93(c)):

$$q^2 = 120^2 - (120 \times 45) - (120 \times 90) + (45 \times 90)$$
$$= 2250$$
$$\underline{q = \pm 47 \cdot 4 \text{ N/mm}^2}$$

and from equation (6):

$$f_{p1} = 120 = \frac{1}{2}[(45 + 90) + \sqrt{(f_y - f_x)^2 + 4q^2}]$$
$$\therefore \sqrt{(f_y - f_x)^2 + 4q^2} = 120 \times 2 - (45 + 90)$$
$$= 105$$
$$\therefore \qquad f_{p2} = \frac{1}{2}[(45 + 90) - 105]$$

minimum principal stress $f_{p2} = 15 \text{ N/mm}^2$

Normally this last part is best solved from first principles.

*Note:*
$$\tan 2\theta = \frac{2 \times 47 \cdot 4}{45 - 90}$$
$$= -2 \cdot 1$$
$$[\tan^{-1}(-2 \cdot 1) = 180° - 64° \, 32']$$
$$2\theta = 180° - 64° \, 32'$$
$$\theta = 57° \, 44'$$

Maximum shear stress $= \frac{1}{2}\sqrt{(45 - 90)^2 + 4 \times 2250}$
$= 52 \cdot 5 \text{ N/mm}^2$ acting at $57° \, 44' - 45° = 12° \, 44'$ to the horizontal,

i.e. for the stress pattern on this material the maximum principal stress is 120 N/mm² compression acting on a plane at $57° \, 44'$ to the horizontal. The maximum shear stress is 52·5 N/mm² acting on a plane at 12° 44′ to the horizontal. The minimum principal stress is 15 N/mm² compression. These are shown on Fig. 93(d).

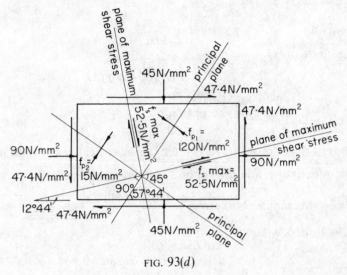

FIG. 93(d)

SPECIMEN QUESTION 49

A bar of circular cross-section can be subjected to a torque, applied about the longitudinal axis, which produces a shear stress at the outside fibres of 60 N/mm², or an axial pull which produces a tensile stress of 90 N/mm². If the torque and the axial pull are applied simultaneously, determine the principal stresses and the maximum shear stress. Determine also the positions of the planes on which they act, relative to the longitudinal axis of the bar.

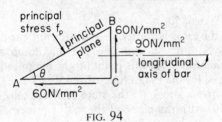

FIG. 94

SOLUTION

The stresses are shown in Fig. 94. Resolve forces horizontally:

$$f_p \text{ AB} \sin \theta = 60 \text{ AC} - 90 \text{ BC}$$
$$f_p \sin \theta = 60 \cos \theta - 90 \sin \theta \qquad (1)$$

Resolve forces vertically:

$$f_p \, AB \cos \theta = 60 \, BC$$
$$f_p \cos \theta = 60 \sin \theta \qquad\qquad (2)$$

from (2)

$$\sin \theta = \frac{f_p \cos \theta}{60}$$

substitute in (1):

$$f_p{}^2 \frac{\cos \theta}{60} = 60 \cos \theta - 90 \frac{f_p \cos \theta}{60}$$
$$f_p{}^2 + 90 f_p - 3600 = 0$$
$$(f_p + 120)(f_p - 30) = 0$$
$$f_p = -120 \quad \text{or} \quad +30$$

The principal stresses are 120 N/mm$^2$ tension (negative sign indicates opposite to direction shown) or 30 N/mm$^2$ compression.

$$\text{Maximum shear stress} = \frac{120 - (-30)}{2} = 75 \text{ N/mm}^2$$

Also, from equation (2),

$$\tan \theta = f_p / 60$$
$$= -120/60 \quad \text{or} \quad 30/60$$
$$= -2 \quad \text{or} \quad \tfrac{1}{2}$$

For 120 N/mm$^2$ tensile stress, $\theta_1 = \tan^{-1}(-2) = (180° - 63° \, 26') = \underline{116° \, 34'}$ to the longitudinal axis.

For 30 N/mm$^2$ compressive stress, $\theta_2 = \tan^{-1}(-\tfrac{1}{2}) = \underline{26° \, 34'}$ to the longitudinal axis.

For maximum shear stress, angle with longitudinal axis = 116° 34' − 45° = $\underline{71° \, 34'}$.

and a complementary maximum shear stress at 90° + (71° 34') = $\underline{161° \, 34'}$ to the longitudinal axis.

SPECIMEN QUESTION 50

At a point in the cross-section of a loaded beam, the major principal stress is 140 N/mm$^2$ tension and the maximum shear stress is 76 N/mm$^2$. Determine for this point (a) the magnitude of the minor principal stress, (b) the magnitude of the direct stress on the plane of maximum shear stress, (c) the state of stress on a plane making an angle of 30° with the plane of the major principal tensile stress.

SOLUTION

(a) Since the maximum shear stress = half the difference between principal stress:

$$76 = \frac{140 - \text{minimum } f_p}{2}$$

Minimum principal stress $= -12$, or $\underline{12 \text{ N/mm}^2 \text{ compression.}}$

These are shown in Fig. 95($a$).

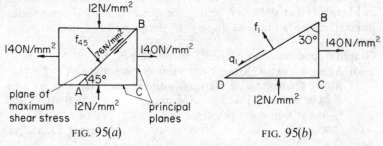

FIG. 95($a$)          FIG. 95($b$)

($b$) For direct stress $f_{45}$ on plane of maximum shear stress, consider part ABC only. Resolve forces at right angles to AB:

$$f_{45}\text{AB} = 12 \text{ AC} \cos 45° - 140 \text{ BC} \sin 45°$$
$$f_{45} = 12 \cos^2 45° - 140 \sin^2 45°$$
$$= 12 \times 0{\cdot}5 - 140 \times 0{\cdot}5 = -64$$

or $\underline{\underline{64 \text{ N/mm}^2}}$ tension (minus sign indicates opposite direction shown).

($c$) A plane making an angle of 30° with the plane of the major principal tensile stress is shown as plane DB (Fig. 95($b$)). The direct stress and shear stress on this plane are $f_1$ and $q_1$ respectively.

Resolve forces at right angles to DB:

$$f_1 \text{ DB} = 140 \text{ BC} \cos 30° - 12 \text{ DC} \sin 30°$$
$$f_1 = 140 \cos^2 30° - 12 \sin^2 30°$$
$$= 140 \times 0{\cdot}75 - 12 \times 0{\cdot}25$$
$$= \underline{\underline{102 \text{ N/mm}^2 \text{ tension,}}} \text{ as shown.}$$

Resolve forces along plane DB:

$$q \text{ DB} = 140 \text{ BC} \sin 30° + 12 \text{ DC} \cos 30°$$
$$q = 140 \cos 30° \sin 30° + 12 \sin 30° \cos 30°$$
$$= 140 \times \frac{\sqrt{3}}{2} \times 0{\cdot}5 + 12 \times 0{\cdot}5 \times \frac{\sqrt{3}}{2}$$
$$= \underline{\underline{65{\cdot}8 \text{ N/mm}^2}}$$

The analytical solution always involves resolution of forces in two directions at right angles. There is, however, a graphical method of solving complex stress problems.

## MOHR'S CIRCLE OF STRESS

Mohr's circle of stress provides a simple means of arriving at various stresses by graphical means. The proof of this construction is simple but tedious, and therefore has not been given.

Assume a body is subject to principal stress $f_{p1}$ and $f_{p2}$ as shown in Fig. 96(a). Let stresses $f_1$ and $q_1$ be the direct and shear stresses respectively on a plane at angle $\theta$ to the plane on which stress $f_{p1}$ acts.

Graphical construction proceeds as follows (*see* Fig. 96(b)):

(a) Along an axis OX mark off point A such that $OA = f_{p2}$.
(b) Along the same axis mark off point B such that $OB = f_{p1}$ (it is here assumed that $f_{p1}$ is the maximum principal stress and $f_{p2}$ is the minimum principal stress).
(c) Bisect AB at D and draw a semi-circle, centre D, radius DA = DB.
(d) Draw chord AC such that angle CAB = $\theta$.
(e) Drop a perpendicular from C on to OX to cut at E.

Now:

OE = direct stress $f_1$ on plane at angle $\theta$
EC = shear stress $q_1$ on plane at angle $\theta$

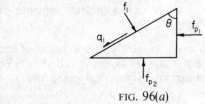

FIG. 96(a)

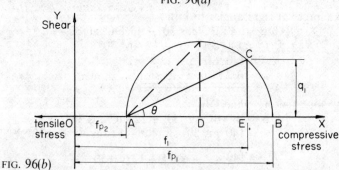

FIG. 96(b)

Also note: the maximum possible value of shear force is when $\theta = 45°$ and its value will be equal to OD, where $OD = f_{p2} + \frac{1}{2}(f_{p1} - f_{p2}) = (f_{p2} - f_{p1})/2$ (as proven).

The principal stresses given should be drawn with compression and tension in opposite directions, i.e. if $f_{p2}$ were tensile and $f_{p1}$ compressive, A would be to the left of 0 and B to the right. Otherwise the construction is as shown.

If two direct stresses at right angles, $f_x$ and $f_y$ are given, together with shearing stress $q$ on the same plane (*see* Fig. 97($a$)) then the construction of circle of stress is as follows:

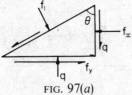

FIG. 97($a$)

(*a*) Along axis OX mark off point $P_1$ such that $OP_1 = f_y$.

(*b*) Along axis OX mark off point $P_2$ such that $OP_2 = f_x$.

(*c*) From $P_2$ set off at right-angles line $P_2C_1$ such that $P_2C_1 = q$.

(*d*) Bisect $P_1P_2$ at D and draw a circle centre D radius $DC_1$ (*see* Fig. 97($b$)).

This gives the Mohr circle of stress (Fig. 97($b$)); OA and OB are the

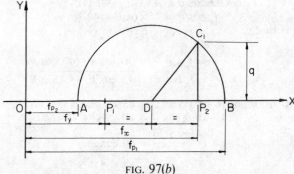

FIG. 97($b$)

minimum and maximum principal stresses respectively. The stresses on any other plane can be found as in the previous construction.

SPECIMEN QUESTION 51

At a point in a piece of material there are two planes at right angles on which there are normal stresses of 90 N/mm² tension and 45 N/mm² tension, together with shearing stresses of 40 N/mm², as indicated in Fig. 98($a$). There are no stresses in the third mutually perpendicular plane.

(*a*) Determine by calculation the principal stresses at the point and the inclinations of the planes on which they act.

(*b*) Verify the results of (*a*) by means of the Mohr stress circle.

SOLUTION

(*a*) Let the principal plane make angle $\theta$ with the horizontal, as shown in Fig. 98($b$). Resolve forces horizontally:
$$f_p \ CD \sin \theta = 90 \ DE - 40 \ CE$$
$$f_p \sin \theta = 90 \sin \theta - 40 \cos \theta \qquad (1)$$

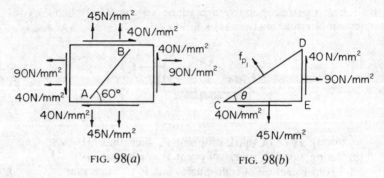

FIG. 98(a)           FIG. 98(b)

Resolve forces vertically:
$$f_p \text{ CD} \cos \theta = 45 \text{ CE} - 40 \text{ DE}$$
$$f_p \cos \theta = 45 \cos \theta - 40 \sin \theta \qquad (2)$$

Divide (1) by (2):

$$\frac{f_p \sin \theta}{f_p \cos \theta} = \frac{90 \sin \theta - 40 \cos \theta}{45 \cos \theta - 40 \sin \theta}$$

$$90 \sin \theta \cos \theta - 40 \cos^2 \theta = 45 \sin \theta \cos \theta - 40 \sin^2 \theta$$
$$45 \sin \theta \cos \theta = 40 (\cos^2 \theta - \sin^2 \theta)$$
$$\frac{45 \sin 2\theta}{2} = 40 \cos 2\theta$$

$$\tan 2\theta = \frac{40 \times 2}{45} = 1{\cdot}78$$
$$2\theta = 60° \ 40'$$
$$\theta = \underline{\underline{30° \ 20'}}$$

From equation (1):
$$f_p = 90 - 40 \cot \theta$$
$$= 90 - 40 \times 1{\cdot}7$$
$$= \underline{\underline{22 \text{ N/mm}^2 \text{ tension}}}$$

(minimum principal stress) at 30° 20′ to horizontal.
Or
$$f_p = 90 - 40 \times \cot(90° + 30° \ 20')$$
$$= 90 + 40 \tan 30° \ 20'$$
$$= 90 + 40 \times 0{\cdot}58$$
$$= \underline{\underline{113{\cdot}2 \text{ N/mm}^2 \text{ tension}}}$$

(maximum principal stress) at 120° 20′ to horizontal.

(b) Construct the stress circle (Fig. 98(c)).

This agrees with the principal stresses calculated in (a) above.

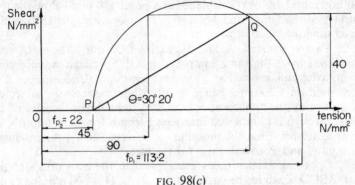

FIG. 98(c)

## EXAMINATION QUESTIONS

1. Illustrate the shear stress distribution in the beam section shown in Fig. 99 for a shearing load of 60 kN. Compare the maximum shear stress with the mean value as usually computed.

2. Calculate the maximum horizontal shear in the beam shown in Fig. 100 where the shearing force is 150 kN. Draw the shear stress distribution diagram for the section. Assume ⊥ section is made up of rectangles.

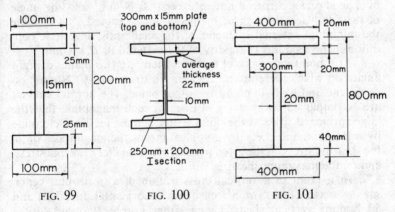

FIG. 99          FIG. 100          FIG. 101

3. Sketch the shear stress variation diagram for the section shown in Fig. 101, noting all variations.

4. A rolled steel beam of ⊥ section is 500 mm deep and 200 mm wide, with flanges 25 mm thick and web 15 mm thick. The cross section may be assumed to be built up of rectangles. The beam is simply supported over a span of six metres and carries a uniformly distributed load of $w$ kN/m on the whole span. If the shearing stresses

on horizontal and vertical planes at a point 150 mm below the top surface of the flange and 1·5 metres from one end are 15 N/mm², determine the value of w.

5. Two specimens, A and B, each of 20 mm diameter, were machined from the same sample of metal. Specimen A, subjected to an axial pull, showed an average extension of 0·036 mm per kN, measured on a 200 mm gauge length, while specimen B, subjected to a pure torque applied about the longitudinal axis, was found to twist through 0·1° in a 200 mm gauge length for every mm kN of applied torque. Find the modulus of direct elasticity, the modulus of rigidity and Poisson's ratio for the metal.

6. Opposing axial torques are applied to the ends of a straight bar, ABCD. Each of the parts AB, BC and CD is 500 mm long and has a hollow circular cross-section, the inside and outside diameters being, respectively, AB 24 mm and 56 mm, BC 24 mm and 64 mm, CD 48 mm and 64 mm. The modulus of rigidity of the material is 90 kN/mm² throughout. Calculate:

   (a) the maximum torque which can be applied, if the maximum shear stress is not to exceed 80 N/mm²;
   (b) the maximum torque if the twist of D relative to A is not to exceed 2°.

7. A hollow shaft in which the external and internal diameters are in the ratio of 5:3 is required to transmit a torque of 57 kN m. The shearing stress is not to exceed 60 N/mm² and the angle of twist in a length of three metres is not to exceed 1°. Calculate the minimum external diameter of the shaft satisfying these conditions. The modulus of rigidity of the material is 82 kN/mm².

8. At a point in a piece of material there are two planes at right angles on which there are normal tensile stresses of 80 N/mm² on one plane and 45 N/mm² on the other plane. The normal stresses are accompanied by shearing stresses of such magnitude that the larger principal stress at the point is 104 N/mm² tensile. Determine by calculation and check by graphical means the magnitudes of (a) the shearing stress on the given planes, (b) the smaller principal stress, and (c) the maximum shearing stress.

9. At a point in a vertical cross-section of a horizontal beam, stresses occur of 90 N/mm² tensile (in a horizontal direction) and 60 N/mm² vertical shear. Using either graphical or analytical methods, determine for this point (a) the maximum tensile stress, (b) the maximum compressive stress, (c) the maximum shear stress, and (d) the positions of the planes on which these stresses act, relative to the horizontal.

10. The rectangular element shown in Fig. 102 has direct stresses $f_x$ and $f_y$ and shear stresses q acting on the planes shown. If failure occurs when the shear stress on any plane exceeds 34 N/mm², deter-

mine the maximum tensile permissible value of $f_x$ for each of the following cases:

   (a) $f_y = 0$; $q = 0$
   (b) $f_y = 28$ N/mm² compression, $q = 0$
   (c) $f_y = 28$ N/mm² compression, $q = 20$ N/mm².

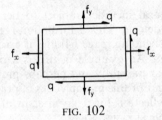

FIG. 102

11. A rectangular block of material is subjected to a horizontal tensile stress of 110 N/mm² in one plane and a tensile stress of 46 N/mm² on a plane at right angles, together with shear stresses of 60 N/mm² in the same directions. Find, by calculation, the magnitude of the direct and shear stresses on a plane at 40° to the direction of the 46 N/mm² stress. Find also the directions of the principal planes, the principal stresses and the maximum shear stress in this block of material. Check all results by graphical means.

# Chapter 5

# SLOPE AND DEFLECTION

ALTHOUGH a structural member subject to bending may not be overstressed, it will deform. This deformation is assumed to be circular (*see* theory of bending, page 64) and if excessive may lead to unsatisfactory design. An example of this is a beam which, when deflected, may appear unsightly and perhaps prevent a door under its span from closing. For this reason codes of practice specify maximum permissible deflection for a given span of beam, as well as maximum permissible stresses.

As mentioned in Chapter 4 (page 94), there is a direct relationship between the load, the shear force, the bending moment, the slope of the beam, and its deflection at any given point. There are several methods of calculating this slope and deflection, but for simple cases Mohr's theorems may be used directly. These two theorems form the basis of the greater part of advanced structural analysis; any student intending to take this subject further should be sure that he thoroughly grasps their meaning and how to use them. The proof is given here for reference.

## MOHR'S THEOREMS

### First theorem

Consider a length of an originally straight beam. If load is applied, the beam will deflect. The centre line of the deflected beam is shown in Fig. 103(*a*), where points on the centre line of the beam A and B take positions A′ and B′. Fig. 103(*b*) shows the bending moment diagram over this length of beam.

Let the average bending moment over a short portion PQ of the beam, length $\delta x$, distance $x$ from point B = $M$.

Draw tangents to the deflected beam at P and Q and let the angle between these tangents (the change in slope) = $\delta\theta$.

If length of arc PQ = $\delta s$, the angle subtended at its centre will be $\delta\theta$ and the radius of the deflected centre-line is $R$.

$$\delta s = R\delta\theta \quad \text{or} \quad R = \delta s/\delta\theta$$

From the theory of bending:

$$\frac{M}{I} = \frac{f}{y} = \frac{E}{R}$$

or

$$\frac{M}{EI} = \frac{1}{R}$$

124

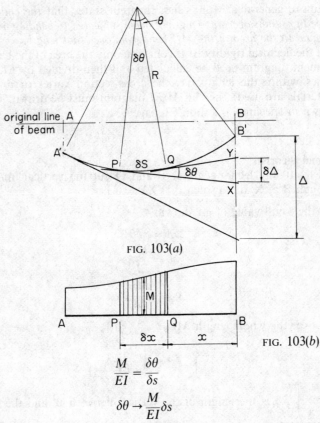

FIG. 103(a)

FIG. 103(b)

$$\therefore \quad \frac{M}{EI} = \frac{\delta\theta}{\delta s}$$

$$\delta\theta \rightarrow \frac{M}{EI}\delta s$$

But for the small deflection over the short length considered:

$$\delta s = \delta x$$

$$\therefore \quad \delta\theta = \frac{M}{EI}\delta x$$

Over the whole length AB:

$$\int_0^\theta \delta\theta = \int_A^B \frac{M}{EI}\delta x$$

where $\int_0^\theta \delta\theta$ is the change of slope from A′ to B′ and $\int_A^B M\delta x$ is the area of the bending moment diagram between A and B.

Assuming $EI$ constant:

the change in slope from A′ to B′

$$= \frac{\text{the area of the bending moment diagram between A and B}}{EI}$$

or, more generally, Mohr's first theorem states that *the change in slope in radians over any length of beam subjected to bending moment is equal to the area of the M/EI diagram over that length.*

If the flectural rigidity *EI* is constant, only the area of the bending moment diagram need be calculated and then divided by *EI*. If *EI* varies within the length (i.e. the cross-section varies or different materials are used), then an *M/EI* diagram must be drawn. This is shown in specimen question 53, page 137.

### Second theorem

In Fig. 103(*a*), let tangents at P and Q cut the vertical intercept through B at X and Y such that XY = $\delta\Delta$.

For the small value of $\delta\Delta$ shown:

$$\delta\Delta = x\delta\theta$$

but

$$\delta\theta = \frac{M}{EI}\delta x$$

$$\therefore \quad \delta\Delta = \frac{Mx}{EI}\delta x$$

and over the whole length AB:

$$\int_0^\Delta \delta\Delta = \int_A^B \frac{Mx}{EI}\delta x$$

where $\int_0^\Delta \delta\Delta$ is the length of the vertical between B' and the point of interception of a tangent from A' with this vertical through B' (*see* Fig. 103(*a*)). $\int_A^B Mx\delta x$ is the first moment of the bending moment diagram between A and B about B.

Thus Mohr's second theorem states that *for an originally straight beam subject to bending moment, the vertical between one terminal and the tangent from the other terminal is equal to the first moment of the M/EI diagram about the terminal where the intercept is measured.* As in the first theorem, if *EI* is constant only the bending moment diagram need be drawn. If *EI* varies, then the *M/EI* diagram is required.

The student should notice that in the general case $\Delta$ is *not* the deflection of the beam. However, it is possible to apply the second theorem to find the value of maximum deflection, when the point of maximum deflection is known, since at this point a tangent to the beam will be horizontal.

EXAMPLE
Find the maximum deflection for a simply supported beam, span $L$, carrying a uniformly distributed load of $w$/unit length across the whole span (*see* Fig. 104($a$)).

Owing to the symmetry of loading, the point of maximum defection will be at the centre, C (Fig. 104($b$)).

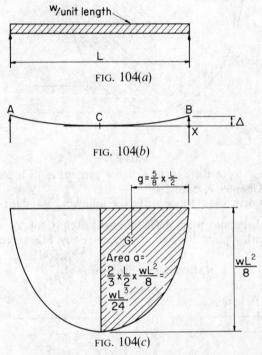

FIG. 104($a$)

FIG. 104($b$)

FIG. 104($c$)

A tangent at C will therefore be horizontal and cut the right-hand support at X (Fig. 104($b$)). Point B is supported and will not deflect; therefore $\Delta_{CB} = BX$, which is also the maximum deflection of the beam.

$$\therefore \quad \Delta_{CB} = \text{maximum deflection} = \frac{\text{area } a \times g}{EI}(\text{see Fig. 104}(c))$$

$$= \frac{wL^3/24 \times 5L/16}{EI} = \frac{5wL^4}{384\,EI}$$

## Properties of a parabola

It is evident from the above example that the area and centroid of a parabola should be readily calculated. The equation of a parabola is $y = mx^2$, which gives a curve as shown in Fig. 105.

The area enclosed by a parabola and the $y$ axis equals $\frac{2}{3}xy$, and the centroid of this area lies $\frac{3}{8}x$ from the $y$ axis. The area enclosed by a parabola and the $x$ axis equals $\frac{1}{3}xy$, and the centroid of this

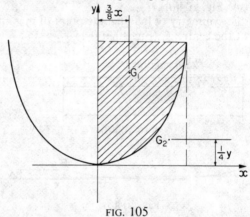

FIG. 105

area lies $\frac{1}{4}y$ from the $x$ axis. Further properties of a parabola are given in Chapter 6, page 160.

The student should now solve the following standard cases:

1. Find the maximum deflection for a simply supported beam of span $L$ with a point load $W$ at the centre (*see* Fig. 106(*a*) and (*b*)).

$$\left(\text{Maximum deflection} = \frac{WL^3}{48EI}\right)$$

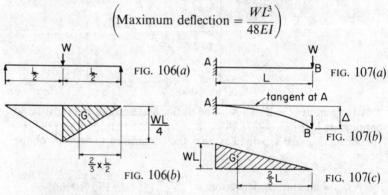

FIG. 106(*a*)

FIG. 107(*a*)

FIG. 107(*b*)

FIG. 106(*b*)

FIG. 107(*c*)

2. Find the maximum deflection for a cantilever, length $L$, carrying a point load $W$ at the end (*see* Fig. 107(*a*), (*b*) and (*c*)).

In this case it is at the support that the beam remains horizontal and $\Delta$ is measured at the point of maximum deflection, i.e. take area-moments about B.

$$\left(\text{Maximum deflection} = \frac{WL^3}{3EI}\right)$$

3. Find the maximum deflection for a cantilever, length $L$, carrying a uniformly distributed load of $w$/unit length along the whole length.

$$\left(\text{Maximum deflection} = \frac{wL^4}{8EI}\right)$$

## Fixed-ended beams

These theorems may also be applied to fixed-ended beams.

EXAMPLE

Find the end moments and maximum deflection for a fixed-ended beam carrying a uniformly distributed load of $w$/unit length over span $L$ (*see* Fig. 108).

(a) *Loaded beam:* in this case, owing to the symmetry of loading, fixing moments $M_A$ and $M_B$ must be equal.

(b) *Fixed bending moment diagram:* total area = $-M_A L$ or $-M_B L$

(c) *Free bending moment diagram:* total area of the "free" bending moment diagram = $\frac{2}{3}\frac{wL^2}{8}L = \frac{wL^3}{12}$

(d) *Final bending moment diagram:* these diagrams may be combined to give the final bending moment diagram. However, in area-moment problems it is often more convenient to keep them separate.

(e) *Deflected beam:* maximum deflection is at centre C.

Since the beam is horizontal at both ends, the difference of slope from A to B is zero, and therefore, from Mohr's first theorem, the area of the bending moment diagram must be zero. This means that the area of the positive "free' bending moment diagram must equal the area of the negative "fixed" bending moment diagram.

$$\frac{wL^3}{12} = M_A L$$

$$M_A = \frac{wL^2}{12}$$

$$M_B = \frac{wL^2}{12}$$

The final moment at the centre of the beam $M_c$

$$= \frac{wL^2}{8} - \frac{wL^2}{12}$$

$$= \frac{wL^2}{24}$$

Also the beam will be horizontal at the point of maximum deflec-

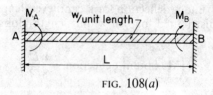

FIG. 108(a)

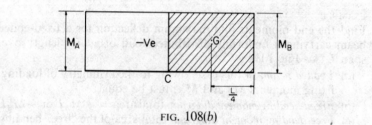

FIG. 108(b)

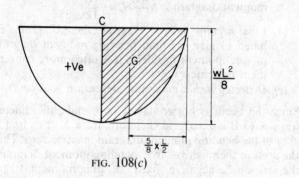

FIG. 108(c)

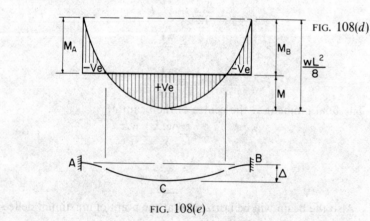

FIG. 108(d)

FIG. 108(e)

tion C. Therefore the first moment of the $M/EI$ diagram from C to B about B will equal the maximum deflection.

It is easier to consider the "fixed" and "free" bending moment diagrams separately, i.e.

First moment of "fixed" bending moment diagram from C to B about B

$$= \left(-M_A \frac{L}{2}\right) \times \frac{L}{4}$$

$$= \left(-\frac{wL^2}{12} \frac{L}{2}\right) \times \frac{L}{4}$$

$$= -\frac{wL^4}{96}$$

First moment of "free" bending moment diagram from C to B about B

$$= \left(\frac{2}{3} \frac{wL^2}{8} \frac{L}{2}\right) \times \frac{5}{8} \times \frac{L}{2}$$

$$= \frac{5wL^4}{384}$$

$$\therefore \quad \text{maximum deflection} = \left(-\frac{wL^4}{96} + \frac{5wL^4}{384}\right) \Big/ EI$$

$$= \frac{wL^4}{384EI}$$

The student should now find the bending moments and maximum deflection for a fixed-ended beam, span $L$, with a point load $W$ at the centre

$$\left(M_A = M_B = \frac{WL}{8} \, M_{\text{centre}} = \frac{WL}{8} \text{ maximum deflection} = \frac{WL^3}{192EI}\right)$$

If the point load $W$ is not at the centre, it is not possible to find the point of maximum deflection by inspection, and hence the maximum deflection cannot be readily found. However, the fixed end moments may be found as follows.

EXAMPLE

Find the fixed end moments for the beam shown in Fig. 109(a).

(a) *Loaded beam:* the fixing moments $M_A$ and $M_B$ will not be equal in this case.

(b) *Fixed bending moment diagram:* area $= -\frac{1}{2}(M_A + M_B)L$

(c) *Free bending moment diagram:* area $= \frac{1}{2}\frac{LWab}{L} = \frac{1}{2}Wab$

(d) *Final bending moment diagram:* the combined diagram may be drawn as shown, but is not convenient for area-moment problems.

(e) *Deflected beam:* note that the bending moment is on the "tension side" of the beam.

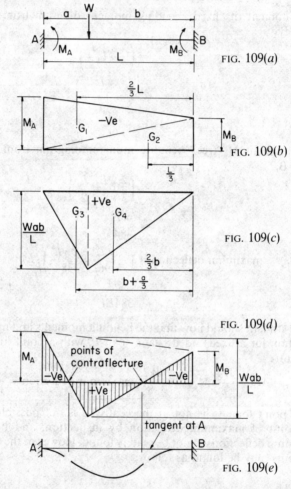

FIG. 109(a)

FIG. 109(b)

FIG. 109(c)

FIG. 109(d)

FIG. 109(e)

For a fixed-ended beam the difference in the slope between the ends A and B is zero, hence the area of the positive bending moment diagram must equal the area of the negative bending moment diagram.

$$\tfrac{1}{2}(M_A + M_B)L = \tfrac{1}{2}Wab$$

$$M_A + M_B = \frac{Wab}{L} \tag{1}$$

As the supports remain on the same level, a tangent through A must pass through B, i.e. $\Delta$ from A to B is zero. Therefore from Mohr's second theorem the moment of the whole bending moment diagram about either support must be zero.

First moment of the "fixed" bending moment diagram about B

$$= (-\tfrac{1}{2}M_A L) \times \frac{2L}{3} + (-\tfrac{1}{2}M_B L) \times \frac{L}{3}$$

$$= -\frac{L^2}{6}(2M_A + M_B)$$

First moment of the "free" bending moment diagram about B

$$= \frac{\tfrac{1}{2}Wab^2}{L} \times \tfrac{2}{3}b + \tfrac{1}{2}\frac{Wa^2b}{L} \times \left(b + \frac{a}{3}\right)$$

$$= \frac{Wab}{6L}(2b^2 + 3ab + a^2)$$

Substitute $a + b = L$ or $a = (L - b)$

$$= \frac{Wab}{6L}[2b^2 + 3b(L - b) + (L - b)^2]$$

$$= \frac{Wab}{6}(b + L)$$

Since the total area moment about B must be zero:

$$-(M_B + 2M_A)\frac{L^2}{6} + \frac{Wab}{6}(b + L) = 0$$

or

$$M_B + 2M_A = \frac{Wab(b + L)}{L^2} \qquad (2)$$

Also, from (1):

$$M_B + M_A = \frac{Wab}{L}$$

Subtracting:

$$M_A = \frac{Wab}{L}\left(\frac{b + L}{L} - 1\right)$$

$$= \frac{Wab^2}{L^2}$$

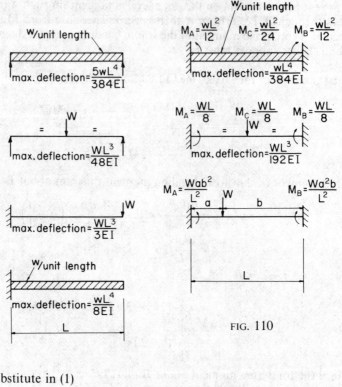

FIG. 110

Substitute in (1)

$$M_B = \frac{Wab}{L} - \frac{Wab^2}{L^2}$$

$$= \frac{Wa^2b}{L^2}$$

All the results for moments and deflections found so far in this chapter are standard cases and should be remembered. Fig. 110 is a summary of these results.

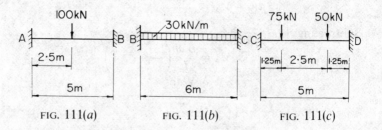

FIG. 111(a)         FIG. 111(b)         FIG. 111(c)

EXAMPLE
Find the fixed end moments for the cases shown in Fig. 111.

ANSWERS
(a) $M_A = M_B = 62 \cdot 5$ kN m
(b) $M_B = M_C = 90$ kN m
(c) $M_C = 64 \cdot 2$ kN m $\quad M_D = 52 \cdot 5$ kN m

SPECIMEN QUESTION 52
Draw the bending moment and shear force diagrams for a fixed-ended beam of four metres span with a 160 kN point load one metre from the left-hand end.

SOLUTION
See the load diagram, Fig. 112(a).

$$\text{Fixed end moments} \quad M_A = \frac{160 \times 1 \times 3^2}{4^2} = 90 \text{ kN m}$$

$$M_B = \frac{160 \times 1^2 \times 3}{4^2} = 30 \text{ kN m}$$

$$\text{"Free" moment under load} = \frac{160 \times 3 \times 1}{4} = 120 \text{ kN m}$$

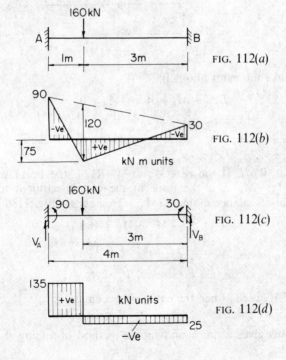

FIG. 112(a)

FIG. 112(b)

FIG. 112(c)

FIG. 112(d)

Final moment under load $= 30 + \frac{3}{4}(90 - 30) = 75$ kN m

These are shown in Fig. 112(b).

To calculate the end reactions in a fixed-ended beam, take moments about one support in the usual manner, taking into account the fixing moments (Fig. 112(c)).

$$\overset{\curvearrowright}{4V_\text{A}} - \overset{\curvearrowright}{90} - \overset{\curvearrowright}{160 \times 3} + \overset{\curvearrowright}{30} = 0$$
$$\underline{V_\text{A} = 135 \text{ kN}}$$
$$\overline{V_\text{B} = 160 - 135 = \underline{\underline{25 \text{ kN}}}}$$

giving the shear force diagram (Fig. 112(d)).

The general case for end reaction may be shown as in Fig. 113(a).

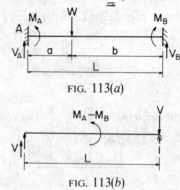

FIG. 113(a)

FIG. 113(b)

Take moments about B:

$$LV_\text{A} - M_\text{A} - Wb + M_\text{B} = 0$$
$$LV_\text{A} = Wb + (M_\text{A} - M_\text{B})$$
$$V_\text{A} = \frac{Wb}{L} + \frac{M_\text{A} - M_\text{B}L}{L}$$

but $Wb/L$ is the reaction at A $(R_\text{A})$ if the beam were not fixed $(M_\text{A} - M_\text{B})/L$ is the force in the couple required to balance the out-of-balance moment $(M_\text{A} - M_\text{B})$, i.e. *see* Fig. 113(b).

$$M_\text{A} - M_\text{B} = V \times L$$
$$V = \frac{M_\text{A} - M_\text{B}}{L}$$
$$\therefore \qquad V_\text{A} = R_\text{A} + V$$

If $M_\text{B} > M_\text{A}$, then the equation becomes:

$$V_\text{A} = R_\text{A} - V$$

This gives a quick and simple method of finding shear force in a fixed beam, but it is important to realise that it is the same as taking moments about a support.

## Beams of varying section

If the moment of inertia of the beam varies along its length, then the $M/I$ diagram should be used for area moment problems.

SPECIMEN QUESTION 53

A horizontal beam ACB, three metres long, is fixed at its ends A and B, which are at the same level. The member has a change of section at its mid-point C such that the second moments of area are $I$ for distance AC and $2I$ for distance CB. A single vertical concentrated load of 320 kN acts at the mid-point C. Determine the fixing moments at A and B.

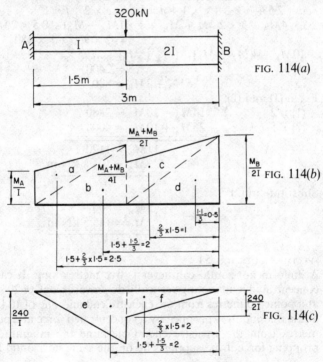

FIG. 114(a)

FIG. 114(b)

FIG. 114(c)

SOLUTION

The beam is shown in Fig. 114(a). The fixed and free $M/I$ diagrams are constructed as in Figs. 114(b) and (c). These are bending moment diagrams but with the ordinate divided by the $I$ value of the mean at each point.

Since there is no change of slope from A to B, the areas of the fixed and free $M/I$ diagrams must be equal ($E$ assumed constant), or:

$$\text{area } a + \text{area } b + \text{area } c + \text{area } d = \text{area } e + \text{area } f$$

$$\tfrac{1}{2}\frac{M_A}{I}1\cdot5 + \tfrac{1}{2}\frac{(M_A + M_B)}{2I}1\cdot5 + \tfrac{1}{2}\frac{(M_A + M_B)}{4I}1\cdot5 + \tfrac{1}{2}\frac{M_B}{2I}1\cdot5$$

$$= \tfrac{1}{2}\frac{240}{I}1\cdot5 + \tfrac{1}{2}\frac{240}{2I}1\cdot5$$

$$4M_A + 2(M_A + M_B) + (M_A + M_B) + 2M_B = 960 + 480$$
$$7M_A + 5M_B = 1440 \qquad (1)$$

Also, a tangent through A passes through B. Take area moments about B (areas are denoted by letters only in the first line):
$$a \times 2\cdot5 + b \times 2 + c \times 1 + d \times 0\cdot5 = e \times 2 + f \times 1$$
$$2\cdot5 \times 4M_A + 2 \times 2(M_A + M_B) + 1 \times (M_A + M_B) + 0\cdot5 \times 2M_B$$
$$= 2 \times 960 + 1 \times 480$$
$$10M_A + 4(M_A + M_B) + (M_A + M_B) + M_B$$
$$= 2400$$
$$5M_A + 2M_B = 800 \qquad (2)$$

From (1) and (2):

(1) × 2  $\qquad 14M_A + 10M_B = 2880$
(2) × 5  $\qquad \underline{25M_A + 10M_B = 4000}$

$$11M_A = 1120$$
$$M_A = 101\cdot8 \text{ kN m}$$

Substitute in (2):

$$2M_B = 800 - 5 \times 101\cdot8$$
$$M_B = 145.5 \text{ kN m}$$

SPECIMEN QUESTION 54

A uniform horizontal cantilever is five metres long. It carries two concentrated loads acting vertically downwards, one of 20 kN acting at a point 1·5 metres from the encastre end and one of 10 kN acting at the free end. It is propped by a concentrated force at a point three metres from the encastre end. Determine the magnitude of the propping force if the supports lie on the same horizontal line.

SOLUTION

This beam can be considered as a cantilever propped at B (see Fig. 115(a)). If the value of the prop is taken as $R$, the bending moment diagram for the prop alone is shown in Fig. 115(b). The bending moment diagram for the loaded cantilever is shown in Fig. 115(c).

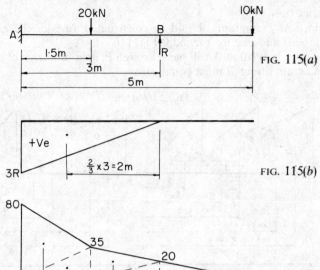

FIG. 115(a)

FIG. 115(b)

FIG. 115(c)

Since a tangent at A will pass through B, and $EI$ is constant, the moment of the total diagram between A and B about B must equal zero, or:

$$\tfrac{1}{2}3R3 \times 2 = \tfrac{1}{2}80 \times 1\cdot5 \times 2\cdot5 + \tfrac{1}{2}35 \times 1\cdot5 \times 2 + \tfrac{1}{2}35 \times 1\cdot5 \times 1$$
$$+ \tfrac{1}{2}20 \times 1\cdot5 \times 0\cdot5$$

$$12R = 200 + 70 + 35 + 10$$
$$R = 315/12 = \underline{\underline{26\cdot25 \text{ kN}}}$$

SPECIMEN QUESTION 55

A horizontal member is fixed at one end and supported by a prop at the other end, as shown in Fig. 116(a). The second moments of area of the three sections are in the ratio $I, 2I, 3I$, the strongest section being at the fixed end. Find the vertical reaction $R$ in the prop if a load of 120 kN is applied at A.

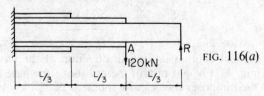

FIG. 116(a)

SOLUTION

Two $M/I$ diagrams should be constructed, one for the prop (Fig. 116(b)) and one for the load (Fig. 116(c)).

As a tangent at A will pass through B, area moments of the whole diagram about B must equal zero.

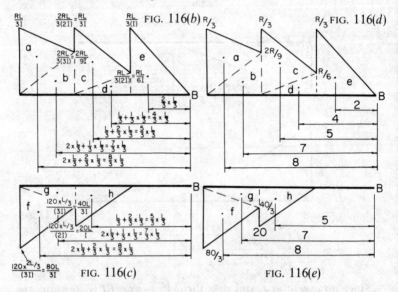

FIG. 116(b)  FIG. 116(d)

FIG. 116(c)  FIG. 116(e)

In determining areas (a) to (h) ($\frac{1}{2}$ base × height), the $\frac{1}{2}$, the $L/3$ and the $L/I$ will occur in each term of the equation and may therefore be cancelled throughout. These areas may then be taken to equal the vertical ordinate of the respective triangles divided by $L/I$.

Also, in the lever arm calculations $L/3$ and $\frac{1}{3}$ may be cancelled throughout. The $M/I$ diagrams can then be represented in the simplified form shown in Fig. 116(d) and (e).

From Figs. 116(d) and (e):

$$8\frac{R}{3} + 7 \times \frac{2R}{9} + \frac{5R}{3} + \frac{4R}{6} + \frac{2R}{3} = \frac{8 \times 80}{3} + \frac{7 \times 40}{3} + 5 \times 20$$

$$\frac{R}{9}(24 + 14 + 15 + 6 + 6) = \frac{20}{3}(32 + 14 + 15)$$

$$R = \frac{3660}{65} = \underline{56\cdot3 \text{ kN}}$$

SPECIMEN QUESTION 56

A uniform vertical column is fixed in direction at the base and is six metres long. At a height of four metres above the base it carries a bracket which imposes a clockwise moment of 120 kN m at this position. Draw the bending moment diagram for the column, indicating on it all important values, (*a* if the upper end be assumed to be pinned in position, (*b*) if the upper end be assumed fully fixed in position and direction.

SOLUTION

(*a*) The bending moment diagrams are shown in Fig. 117(*a*). A tangent at A must pass through the top pin B. From Mohr's second theorem, area moments about B must equal zero:

$$-(\tfrac{1}{2} \times M_A \times 6)(\tfrac{2}{3}6) + (\tfrac{1}{2} \times 80 \times 4)(2 + \tfrac{4}{3}) - (\tfrac{1}{2} \times 40 \times 2)(\tfrac{2}{3} \times 2) = 0$$

$$12M_A = 160 \times \frac{10}{3} - 40 \times \frac{4}{3}$$

$$M_A = \frac{1600 - 160}{3 \times 12}$$

$$= \underline{40 \text{ kN m}} \quad \text{(positive and therefore in the direction}$$

shown).

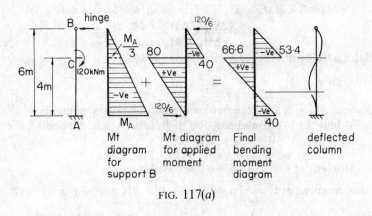

FIG. 117(*a*)

To find the reaction at B, take moments about A:

$$6R_B - 120 = 40$$
$$R_B = 26 \cdot 67 \text{ kN}$$
$$\text{Moment at C} = 26 \cdot 67 \times 2 = \underline{\underline{53 \cdot 34 \text{ kN m}}}$$

$$\text{Also moment at C} = 53 \cdot 34 - 120 = \underline{\underline{-66 \cdot 66 \text{ kN/m}}}$$

(*see* Fig. 117(*a*)).

(*b*) The bending moment diagrams are shown in Fig. 117(*b*). As there is no difference of slope between A and B, the area of the bending moment diagram equals zero, or:

$$6M_B + \tfrac{1}{2}(M_A - M_B)6 = \tfrac{1}{2} \times 80 \times 4 - \tfrac{1}{2} \times 40 \times 2$$
$$3M_A + 3M_B = 160 - 40$$
$$M_A + M_B = 40 \tag{1}$$

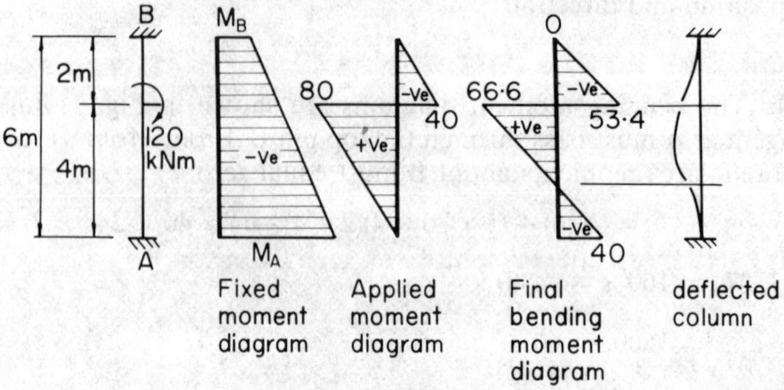

FIG. 117(b)

Also, area moments about B must equal zero:

$$6M_B \times 3 + \tfrac{1}{2}(M_A - M_B)6 \times (\tfrac{2}{3} \times 6) = 160(2 + \tfrac{4}{3}) - 40(\tfrac{2}{3} \times 2)$$
$$12M_A + 6M_B = 480$$
$$2M_A + M_B = 80 \tag{2}$$

Subtract (1) from (2):    $M_A = \underline{\underline{40 \text{ kN m}}}$

Substitute in (1):    $M_B = \underline{\underline{0}}$

*Note:* if $M_B = 0$, then the fixed moment diagram is triangular.

To find the horizontal reaction at B, take moments about A:
$$0 + 6R_B - 120 = 40$$
$$R_B = 26 \cdot 7 \text{ kN}$$
Moment at C $= 0 + 26 \cdot 7 \times 2 = \underline{\underline{53 \cdot 4 \text{ kN m}}}$

Also, moment at C $= 53 \cdot 4 - 120 = \underline{\underline{-66 \cdot 6 \text{ kN m}}}$ *(see* Fig. 117(b)).

*Note:* it can be seen that the moments are the same for both fixed and hinged condition at B.

# DEFLECTION

Mohr's theorems may also be used to determine the deflection of a beam at any given point.

SPECIMEN QUESTION 57

The beam shown in Fig. 118(a) is of uniform flectural rigidity, *EI*. Calculate the deflection at the centre of span AB, in terms of *W*, *L* and *EI*.

SOLUTION

$$R_A = \frac{W \times 3L + W + L/2 - 2W \times L}{2L} = 0.75\ W$$

$$R_B = W + W + 2W - 0.75W \qquad = 3.25W$$

The bending moment diagram is shown in Fig. 118(b) and the deflected beam in Fig. 118(c). The deflected shape can be sketched as the bending moment diagram is entirely negative, i.e. the beam is hogging along its whole length.

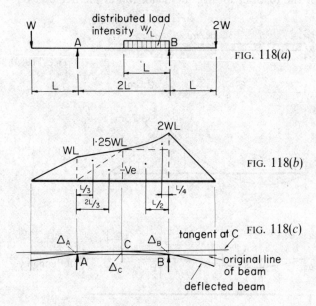

FIG. 118(a)

FIG. 118(b)

FIG. 118(c)

A tangent at the mid-point C gives vertical intercepts at A and B of $\Delta_A$ and $\Delta_B$ respectively.

To find $\Delta_A$, take area moments of the bending moment diagram between A and C about A:

$$\Delta_A = \frac{\frac{1}{2}WL^2 \times \dfrac{L}{3} + \frac{1}{2}1.25WL^2 \times \dfrac{2L}{3}}{EI} = \frac{7WL^3}{12EI}$$

Similarly, find $\Delta_B$:

$$\Delta_B = \frac{\frac{1}{3}0.75WL^2 \times \dfrac{L}{4} + 1.25WL^2 \times \dfrac{L}{2}}{EI} = \frac{8.25WL^3}{12EI}$$

Mid-way between these points give the deflection at C:

$$= \frac{(7 + 8 \cdot 25)}{2} \frac{WL^3}{12} = \frac{61WL^3}{96EI} \quad upwards.$$

SPECIMEN QUESTION 58

Fig. 119(a) shows a horizontal beam ABC of uniform flectural rigidity *EI* kN m² units. The beam is simply supported at A and B. Determine, for the loading shown, the deflection at the free end C.

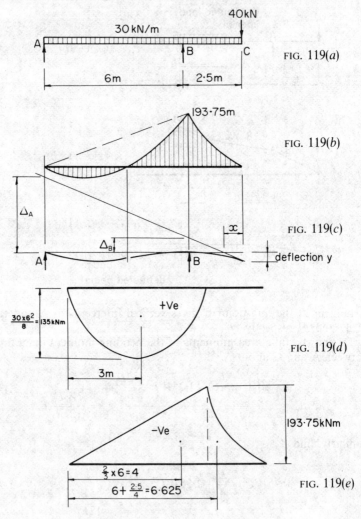

FIG. 119(a)

FIG. 119(b)

FIG. 119(c)

FIG. 119(d)

FIG. 119(e)

SOLUTION

$$R_B = \frac{40 \times 8\cdot 5 + 30 \times 8\cdot 5^2/2}{6} = 237\cdot 3 \text{ kN}$$
$$R_A = 40 + 30 \times 8\cdot 5 - 236\cdot 7 = 58\cdot 3 \text{ kN}$$

The bending moment diagram is as shown in Fig. 119(b). The deflected shape of the beam will then be as shown in Fig. 119(c) (sagging when the bending moment diagram is positive and hogging when the bending moment diagram is negative).

For ease of calculation, the bending moment diagram is best split into the span diagram (Fig. 119(d)) and the cantilever diagram (Fig. 119(e)).

On the deflection diagram (Fig. 117(e)), draw a tangent from C. Then:

$$\Delta_A = \frac{\frac{2}{3} \times 135 \times 6 \times 3 - \frac{1}{2} \times 193\cdot 75 \times 6 \times 4 - \frac{1}{3} \times 193\cdot 75 \times 2\cdot 5 \times 6\cdot 625}{EI}$$

$$= -\frac{1770}{EI} \quad \text{(negative sign indicates upwards)}$$

Also:

$$\Delta_B = -\frac{\frac{1}{3} \times 193\cdot 75 \times 2\cdot 5 \times 2\cdot 5/4}{EI} = \frac{101}{EI}$$

By similar triangles (Fig. 119(c)):

$$\frac{1770}{(8\cdot 5 - x)} = \frac{101}{(2\cdot 5 - x)} = \frac{y}{x} \qquad \frac{1770}{x(8\cdot 5 - x)EI} = \frac{101}{(2\cdot 5 - x)EI} = \frac{y}{x}$$

$$1770(2\cdot 5 - x) = 101(8\cdot 5 - x)$$
$$4425 - 1770x = 858\cdot 5 - 101x$$
$$x = 3566\cdot 5/1669 = 2\cdot 14 \text{ m}$$
$$\therefore \frac{101}{0\cdot 36EI} = \frac{y}{2\cdot 14}$$
$$y = \frac{101 \times 2\cdot 14}{0\cdot 36EI} = \underline{\underline{600/EI \text{ metres}}}$$

This method of determining deflection is suitable when the point where deflection is required is given. If, however, the maximum deflection is asked for, unless the loading is symmetrical the point of maximum deflection is not known. The position of the point of maximum deflection may be found by using the conjugate beam theory. This involves *loading* the original beam with the bending moment diagram and finding the point of zero shear for this loading. The point so found is the point of maximum deflection for the originally loaded beam. Except for very simple cases however, this is a very tedious approach.

The deflection at any point, including maximum deflection, may be found more easily by using Macauley's method of analysis. This method may also be used to find fixed-end moments and the slope of a beam at any point.

## MACAULEY'S METHOD FOR SLOPE AND DEFLECTION

### Relationship between load, shear force, bending moment, slope and deflection

It has been shown in Chapter 4 (page 95) that:

$$w = -\frac{\delta Q}{\delta x} \qquad \text{or} \qquad Q = \int -w\delta x$$

and
$$Q = \frac{\delta M}{\delta x} \qquad \text{or} \qquad M = \int Q\delta x$$

Consider the short length of beam in Fig. 82 when deflected (Fig. 120).

$$\tan \theta = \frac{\delta y}{\delta x} \quad (\theta = \text{slope of the beam at A})$$

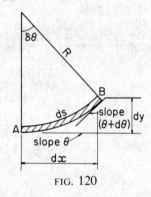

FIG. 120

For small angles,
$$\delta\theta = \tan \delta\theta$$
$$\therefore \qquad \theta = \frac{\delta y}{\delta x} \quad \text{or} \quad y = \int \theta \delta x$$

Since $\delta y$ and $\theta$ are very small, $\delta s$ and $\delta x$ may be taken as equal.
$$\therefore \qquad \delta s = \delta x = R\delta\theta$$
or
$$\frac{1}{R} = \frac{\delta\theta}{\delta x}$$

but
$$\frac{M}{I} = \frac{f}{y} = \frac{E}{R}$$

∴
$$\frac{1}{R} = \frac{M}{EI} = \frac{\delta\theta}{\delta x}$$

or
$$\theta = \frac{1}{EI}\int M\delta x \quad (EI \text{ taken as constant})$$

i.e.          load intensity $= -w$

shear force    $= Q = -\int w\delta x$

moment    $= M = \int Q\delta x$

slope    $= \theta = \frac{1}{EI}\int M\delta x$

deflection    $= y = \int \theta\delta x$

Hence if the load intensity at a point is known, all the other values may be found by successive integration.

It may also be summed up thus:

$$
\begin{aligned}
\text{deflection} &= y \\
\text{slope} &= \theta &&= \delta y/\delta x \\
\text{moment} &= M &&= \delta^2 y/\delta x^2 EI \\
\text{shear force} &= Q &&= \delta^3 y/\delta x^3 EI \\
\text{load intensity} &= -w &&= \delta^4 y/\delta x^4 EI
\end{aligned}
$$

SPECIMEN QUESTION 59

Find expressions for (a) shear force, (b) bending moment, (c) slope and (d) deflection at a point distance $x$ from one support of a simply supported beam, span $L$, carrying a uniformly distributed load $w$.

SOLUTION
See Fig. 121.

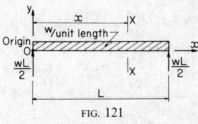

FIG. 121

(a) Shear force $Q = -\int w\delta x$

$$Q = -wx + C_1$$

By inspection of the beam:

when $\quad x = L/2, \quad Q = 0 \quad \therefore \quad C_1 = \dfrac{wL}{2}$

$$Q = -wx + \frac{wL}{2}$$

(b) Moment $\quad M = \displaystyle\int \theta \delta x$

$$M = -\frac{wx^2}{2} + \frac{wLx}{2} + C_2$$

By inspection:

when $\quad x = 0, \quad M = 0 \quad \therefore \quad C_2 = 0$

$$M = \frac{wx^2}{2} + \frac{wLx}{2}$$

Both these expressions could have been written down directly, without the use of integration.

(c) Slope $\quad \theta = \dfrac{1}{EI}\displaystyle\int M \delta x$

$$\theta EI = \frac{wx^3}{6} + \frac{wLx^2}{4} + C_3$$

At the point of maximum deflection, when $x = L/2$, then $\theta = 0$.

$\therefore \quad\quad C_3 = +\dfrac{wL^3}{6 \times 2^3} - \dfrac{wLL^2}{4 \times 2^2} = -\dfrac{wL^3}{24}$

and $\quad\quad \theta EI = -\dfrac{wx^3}{6} + \dfrac{wLx^2}{4} - \dfrac{wL^3}{24}$

(d) Deflection $\quad y = \displaystyle\int \theta \delta x$

$$yEI = -\frac{wx^4}{24} + \frac{wLx^3}{12} - \frac{wL^3 x}{24} + C_4$$

By inspection, when $x = 0, \quad y = 0, \quad \therefore \quad C_4 = 0$

$$yEI = -\frac{wx^4}{24} + \frac{wLx^3}{12} - \frac{wL^3 x}{24}$$

To check deflection, maximum deflection is when $x = L/2$

$$yEI = -\frac{wL^4}{24 \times 2^4} + \frac{wL \times L^3}{12 \times 2^3} - \frac{wL^3 \times L}{24 \times 2}$$

$$y = -\frac{5wL^4}{384EI} \quad \text{(the standard expression).}$$

SPECIMEN QUESTION 60

A beam AB is simply supported at the ends and has a constant flexural rigidity ($EI$). It is loaded as shown in Fig. 122($a$). Calculate, in terms of $EI$ (kN m$^2$ units), the value of the maximum deflection.

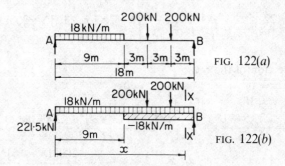

FIG. 122($a$)

FIG. 122($b$)

SOLUTION

Take the origin at A and consider a section XX, distance $x$ from A. Section XX should be taken at a point past the last change of loading on the beam. *Note:* when the uniformly distributed load does not run right across the beam, it is convenient to assume that it does and to take a negative load of the same intensity over the unloaded portions to compensate (*see* Fig. 122($b$)).

Calculate $R_A$ (221·5 kN).

The expression for the bending moment at section XX may now be written down directly:

$$M_{XX} = EI\frac{\delta^2 y}{\delta x^2}$$
$$= 221{\cdot}5x - \frac{18x^2}{2} - \left(\frac{-18[x-9]^2}{2}\right) - 200[x-12] - 200[x-15]$$

(reaction) (+ $V_e$ UDL)     (− $V_e$ UDL)    (200 kN load)    (200 kN load)

If the bending moment were required at some other section than XX, only loads to the left of the section would be considered, i.e. all terms in the above expression would be ignored when the portion inside square brackets was negative. This would also apply to the slope and deflection expressions which follow, therefore the square brackets are retained when integrating.

$$M_{xx} = EI\frac{\delta^2 y}{\delta x^2} = 221 \cdot 5x - 9x^2 + 9[x-9]^2 - 200[x-12]$$

$$- 200[x-15]$$

$$Slope\ \theta_{xx}EI = EI\frac{\delta y}{\delta x} = \frac{221 \cdot 5x^2}{2} - \frac{9x^3}{3} + \frac{9[x-9]^3}{3} - \frac{200[x-12]^2}{2}$$

$$- \frac{200[x-15]^2}{2} + C_1$$

$$EIy = \frac{221 \cdot 5x^3}{6} - \frac{9x^4}{12} + \frac{9[x-9]^4}{12} - \frac{200[x-12]^3}{6} - \frac{200[x-15]^3}{6}$$

$$+ C_1x + C_2$$

By inspection of the beam:

when $x = 0$,  $y = 0$,  $\therefore$  $C_2 = 0$

when $x = 18$,  $y = 0$

$$\therefore\ 0 = \frac{221 \cdot 5 \times 18^3}{6} - \frac{9 \times 18^4}{12} + \frac{9[18-9]^4}{12} - \frac{200[18-12]^3}{6}$$

$$- \frac{200[18-15]^3}{6} + 18C_1$$

$$C_1 = -7410 \cdot 4$$

Maximum deflection occurs when $\theta = 0$. By inspection, $x$ will probably be slightly greater than 9m, or:

$$0 = \frac{221 \cdot 5x^2}{2} - \frac{9x^3}{3} + \frac{9[x-9]^3}{3} - 7410 \cdot 4$$

(all other terms will have negative values inside the square brackets). Multiply by 2 and expand the cubed term:

$$221 \cdot 5x^2 - 6x^3 + 6(x^3 + 243x - 27x^2 - 729) - 14820 \cdot 8 = 0$$

$$59 \cdot 5x^2 + 1458x - 19194 \cdot 8 = 0$$

which is a quadratic equation giving $x = 9 \cdot 5$ metres, i.e. maximum deflection occurs 9·5 metres from A.

*Note:* if $x$ had been taken as *less than* nine metres,

$$0 = \frac{221 \cdot 5x^2}{2} - \frac{9x^3}{3} - 7410 \cdot 4$$

$$= 221 \cdot 5x^2 - 6x^3 - 14820 \cdot 8$$

a cubic equation giving a value of $x$ *greater than* nine metres at the section.

$$EIy = \frac{221 \cdot 5 \times 9 \cdot 5^3}{6} - \frac{9 \times 9 \cdot 5^4}{12} + \frac{9 \times 0 \cdot 5^4}{12} - 9 \cdot 5 \times 7410 \cdot 4$$

$$y = -\frac{44856}{EI}\ \text{metres}$$

SPECIMEN QUESTION 61

A beam ABC is fixed in position and horizontally in direction at A and freely supported at B. It supports loading as indicated in Fig. 123(a). Calculate (a) the fixing moment at A, and (b) the deflection at the centre of span AB, in terms of EI. Supports A and B remain at the same level after loading.

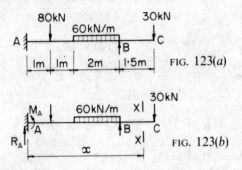

FIG. 123(a)

FIG. 123(b)

SOLUTION

Take A as the origin and select a section XX (see Fig. 123(b)). In this case $R_A$ and $R_B$ cannot be found by simple statics. An expression for $M_{XX}$ can be found in terms of $R_A$, $R_B$ and $M_A$:

$$M_{XX} = EI\frac{\delta^2 y}{\delta x^2} = R_A x - M_A - 80[x-1] - \frac{60[x-2]^2}{2} + \frac{60[x-4]^2}{2} + R_B[x-4]$$

*Note:* a negative uniformly distributed load is taken past B.

$$\theta EI = EI\frac{\delta y}{\delta x} = \frac{R_A x^2}{2} - M_A x - \frac{80[x-1]^2}{2} - \frac{60[x-2]^3}{6} + \frac{60[x-4]^3}{6} + \frac{R_B[x-4]^2}{2} + C_1$$

When $x = 0$, $\theta = 0$, $\therefore$ $C_1 = 0$

$$EIy = \frac{R_A x^3}{6} - \frac{M_A x^2}{2} - \frac{80[x-1]^3}{6} - \frac{60[x-2]^4}{24} + \frac{60[x-4]^4}{24} + \frac{R_B[x-4]^3}{6} + C_2$$

When $x = 0$, $y = 0$, $\therefore$ $C_2 = 0$

When $x = 4$, $y = 0$

$$\therefore \quad 0 = \frac{R_A 4^3}{6} - \frac{M_A 4^2}{2} - \frac{80[4-1]^3}{6} - \frac{60[4-2]^4}{24}$$

which gives

$$4R_A - 3M_A - 150 = 0 \tag{1}$$

When $x = 5.5$,  $M = 0$

$$\therefore \quad 0 = 5.5R_A - M_A - 80[5.5 - 1] - \frac{60[5.5 - 2]^2}{2}$$
$$+ \frac{60[5.5 - 4]^2}{2} + R_B[5.5 - 4]$$

which gives:

$$11R_A - 2M_A - 1320 + 3R_B = 0 \tag{2}$$

also

$$R_B = 80 + 60 \times 2 + 30 - R_A$$
$$= 230 - R_A$$

Substitute in (2):

$$11R_A - 2M_A - 1320 + 3(230 - R_A) = 0$$
$$8R_A - 2M_A - 630 = 0 \tag{3}$$

From equations (1) and (3):

$$8R_A - 6M_A = 300$$
$$\underline{8R_A - 2M_A = 630}$$
$$-4M_A = -330$$
$$\underline{\underline{M_A = 82.5 \text{ kN m}}} \quad \text{(anti-clockwise, as shown).}$$

Substitute in (3):

$$R_A = \frac{630 + 2 \times 82.5}{8} = 99.4 \text{ kN}$$

For deflection at the centre of span AB ($x = 2$):

$$EIy = \frac{99.4 \times 2^3}{2} - \frac{82.5 \times 2^2}{2} - \frac{80[2 - 1]^3}{6}$$

$$\underline{\underline{y = \frac{219.3}{EI} \text{ metres}}}$$

SPECIMEN QUESTION 62

A uniform beam has a clear span of 6 m between the supports, which are fully restrained. The loading on the beam increases linearly from zero at the ends to reach an intensity of 60 kN/m at the centre of the span. Determine the bending moments at the fixed ends, and sketch the shearing force and bending moment diagrams for the beam.

SOLUTION

To obtain an expression for intensity of load at any point distance $x$ from A, assume that the load increases along the whole length of the beam (Fig. 124(*a*)) and then deduct excess load between C and B, i.e. from similar triangles (Fig. 124(*a*)):

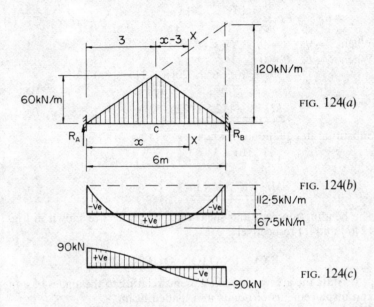

FIG. 124(*a*)

FIG. 124(*b*)

FIG. 124(*c*)

$$w = EI\frac{\delta^4 y}{\delta x^4} = \frac{120}{6}x - \frac{120[x-3]}{3}$$
$$= 20x - 40[x - 3] \qquad (1)$$
$$-Q = EI\frac{\delta^3 y}{\delta x^3} = \frac{20x^2}{2} - \frac{40[x-3]^2}{2} + C_1$$

When $x = 0$, $\quad Q = R_A$ $\quad \therefore \quad C_1 = -R_A$

or:

$$-Q = EI\frac{\delta^3 y}{\delta x^3} = 10x^2 - 20[x - 3]^2 - R_A \qquad (2)$$

For symmetrical beam loading the point of zero shear is at the centre
$\therefore \quad$ when $x = 3$, $\quad Q = 0$
$$0 = 10 \times 3^2 - R_A$$
$$\underline{\underline{R_A = 90 \text{ kN}}}$$
$$-M = EI\frac{\delta^2 y}{\delta x^2} = \frac{10x^3}{3} - \frac{20[x-3]^3}{3} - R_A x + C_2$$

When $x =$ ,   $M = M_A$,   $\therefore$   $C_2 = -M_A$

or:

$$-M = EI\frac{\delta^2 y}{\delta x^2} = \frac{10x^3}{3} - \frac{20[x-3]^3}{3} - R_A x - M_A \qquad (3)$$

$$-\theta EI = EI\frac{\delta y}{\delta x} = \frac{10x^4}{12} - \frac{20[x-3]^4}{12} - \frac{R_A x^2}{2} - M_A x + C_3 \qquad (4)$$

When $x = 0$,   $\theta = 0$,   $\therefore$   $C_3 = 0$

When $x = 6$,   $\theta = 0$

$\therefore$

$$0 = \frac{10 \times 6^4}{12} - \frac{20[6-3]^4}{12} - \frac{90 \times 6^2}{2} - 6M_A$$

$$\underline{\underline{M_A = -112\cdot5 \text{ kN m}}}$$

Substitute in equation (3): when $x = 3$, $M = M_C$

$$-M_C = \frac{10 \times 3^3}{3} - 90 \times 3 + 112\cdot5$$

$$\underline{\underline{M_C = 67\cdot5 \text{ kN m}}}$$

The bending moment and shear force diagrams are shown in Figs. 124(b) and (c) respectively.

## EXAMINATION QUESTIONS

1. State the area–moment theorems relating to the angles of slope and displacements of points in a loaded beam.

A beam of length $L$ and uniform section is pinned to supports at its ends and carries a clockwise moment $M$ at one end. Determine in terms of $M$, $L$, $E$ and $I$ the angles of slope at each end and the position and the value of the maximum deflection.

2. A steel joist of constant flectural rigidity is simply supported at the ends of a 6 m span. The girder carries a uniform load of 16 kN/m over the whole span and a concentrated central vertical downward load of 100 kN. Determine a suitable moment of inertia and depth for the girder if the maximum deflection has not to exceed $\frac{1}{300}$ of the span and the maximum fibre stress due to bending must not exceed 154 N/mm². $E = 210$ kN/mm².

3. A beam of variable stiffness (shown in Fig. 125) carries a load of 250 kN at mid-span. Evaluate the bending moments at the fixed ends of the beam and draw the bending moment diagram, marking on it the appropriate values.

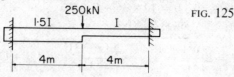

FIG. 125

4. A horizontal cantilever of length $L$ is covered by a uniformly distributed load of intensity $w$ per unit length, and is propped at a point $2L/3$ from the encastre end. Determine the magnitude of the propping force if the supports lie on the same horizontal line.

5. A uniform column 5 m high is fixed at the base and pin-jointed at the upper end. At a point 3 m above the base, a couple of 30 kN m is applied, by a bracket attached to the side of the column. Sketch the shearing force and bending moment diagrams for the stanchion, stating the principal values.

6. A vertical column 8 m high is fixed in position and direction at the base and is hinged at the top. At a height of 6 m above the base it carries a bracket which imposes a clockwise moment of 100 kN m at this position. The second moment of area is also reduced at the same point to one half of the value for the lower 6 m of column. Draw the bending moment diagram for the column, indicating on it all important values.

7. A steel beam of uniform cross-section is simply supported on a span of 8 m, and carries concentrated loads of 20 kN and 60 kN at distances 2 m and 5 m respectively from the left hand end. $I$ for the beam is $100 \times 10^6$ mm$^4$ and $E = 210$ kN/mm$^2$. Determine the deflections at the centre of the beam, and under the 60 kN load.

8. A cantilever 3 m long of constant flectural rigidity $EI$ is rigidly fixed at A. It carries a uniformly distributed load of 6 kN/m for the first 2 m from A, together with point loads of 4·5 kN at B and 2·5 kN

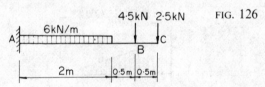

FIG. 126

at the free end C, as shown in Fig. 126. Determine the vertical deflection of points B and C and the rotation at end C, all in terms of $EI$.

9. A cantilever projects 2·5 m from A, where all rotation is prevented. It carries a point load of 20 kN at the free end B, as shown in Fig. 127 (see p. 156). If the second moment of area of cross-section changes along the length of the cantilever as shown, calculate the vertical deflection at B. $E = 210$ kN/mm$^2$.

10. A beam ABC is simply supported at A and B and loaded as shown in Fig. 128 (see p. 156). Calculate, in terms of $EI$, (a) the deflection at the centre of span AB, (b) the deflection at C, (c) the slope at A and B.

11. A horizontal beam, of uniform section and 6 m long, is simply supported at its ends. Two vertical concentrated loads of 50 kN and

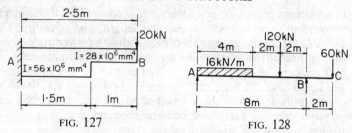

FIG. 127                    FIG. 128

40 kN act 1 m and 3 m respectively from the left-hand support.
Determine the position and magnitude of the maximum deflection,
if $E = 210$ kN/mm$^2$ and $I = 84 \times 10^6$ mm$^4$.

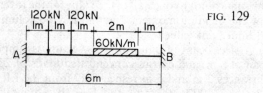

FIG. 129

12. A beam of uniform cross-section is fixed horizontally at the
ends and loaded as shown in Fig. 129. Calculate, by the Macauley
method, (a) the fixed end moments at A and B, (b) the position and
magnitude of maximum deflection in terms of $EI$.

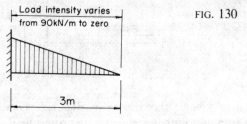

FIG. 130

13. The cantilever shown in Fig. 130 carries a distributed load.
The intensity of the load varies linearly from 90 kN/m at the fixed
end to zero at the free end. Calculate the deflection at the free end,
given that $I = 100 \times 10^6$ mm$^4$ and $E = 210$ kN/mm$^2$.

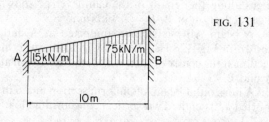

FIG. 131

14. A beam having a span of 10 m and moment of inertia $650 \times 10^6$ mm$^4$ is fixed horizontally at the ends A and B, and loaded with a distributed load varying from 15 kN/m at A to 75 kN/m at B, as shown in Fig. 131. Calculate the reactions and fixing moments at A and B, and the deflection at mid-span. Sketch the bending moment diagram, giving the position and value of maximum sagging moment. $E = 210$ kN/mm$^2$.

## Chapter 6

## ARCHES

### BENDING MOMENTS

CONSIDER a structural member pinned at points A and B, distance $L$ apart, to form an arch, as in Fig. 132(a).

Since the member is restrained in the arch shape there must be horizontal reactions at A and B causing this restraint. If the arch is to be in equilibrium then the horizontal reaction at A must equal that at B. Let this horizontal reaction equal $H$.

When loads $W_1$, $W_2$ and $W_3$ are applied to the arch the vertical reactions $V_A$ and $V_B$ can be found as for a straight beam, provided A and B are at the same level. The bending moment at any point P in the member may also be found, but will be less than that for a straight member by an amount $H \times b$ due to the horizontal reaction, i.e.

$$\text{bending moment at P} = [V_A \times a - W_1(a - l_1)] - H \times b$$
$$= \text{bending moment for horizontal beam}$$
$$- H \times b$$

To draw the bending moment diagram for the arch, first draw that for the straight beam and deduct $H \times b$ from this.

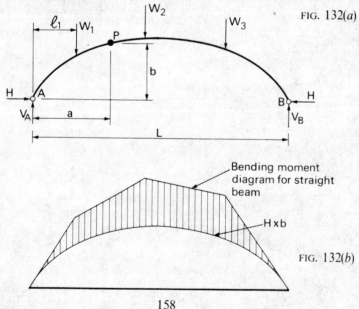

FIG. 132(a)

Bending moment diagram for straight beam

H x b

FIG. 132(b)

158

Since $H$ is constant for any one system of loading, $H \times b$ may be represented by the arch itself, drawn to suitable scale.

The final arch bending moment diagrams then becomes the shaded portion in Fig. 132(b).

## LINE OF THRUST

Consider the line of thrust of resultant reaction $R_A$ shown in Fig. 133(a). When it cuts the line of action of load $W_1$ the resultant of $R_A$ and $W_1$ will be $T_2$. Similarly the resultant of $T_2$ and $W_2$ will be $T_3$, and the resultant of $T_3$ and $W_3$ will be $T_4$. For equilibrium $T_4$ must be equal and opposite to resultant reaction $R_B$.

Consider any point $P$ on the arch rib (*see* Fig. 133(b)).

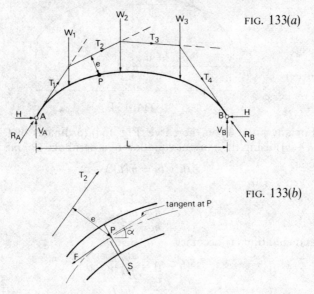

FIG. 133(a)

FIG. 133(b)

The thrust $T_2$ is the resultant of $H$, $V_A$ and $W_1$ and must be resisted by the arch rib. The internal forces caused by thrust $T_2$ will be:

(1) normal axial thrust $F$;
(2) shear resistance $S$.

There will also be a moment of resistance equal to $T_2 \times e$.

The moment of resistance $T_2 \times e$ will be equal to the applied moment at P which as has been shown equals the bending moment for a horizontal beam minus $H \times b$.

To find the normal axial thrust $F$, resolve the forces which make up $T_2$ in a direction along the arch rib:

$$F = H \cos \alpha + V_A \sin \alpha - W_1 \sin \alpha$$

and to find the shear force $S$ resolve the forces which make up $T_2$ in a direction at right angles to the rib:

$$S = V_A \cos \alpha - H \sin \alpha - W_1 \cos \alpha$$

### Geometric properties of a parabolic arch

Arches may have a variety of shapes but many problems deal with a parabolic arch. Some properties of a parabola have been given in Chapter 5, page 127. The properties relevant to the analysis of parabolic arches are given below.

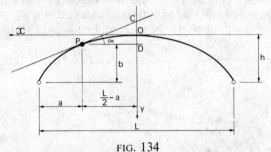

FIG. 134

For any point P on the curve (Fig. 134) (ordinates of P $(L/2 - a)$, $(h - b)$) using the normal equation of a parabola $y = mx^2$,

$$(h - b) = m(L/2 - a)^2 \tag{1}$$

when $a = 0$

$$b = 0$$
$$h = m(L^2/4)$$
$$m = 4h/L^2$$

and equation (1) becomes

$$(h - b) = \frac{4h}{L^2}\left(\frac{L}{2} - a\right)^2$$

$$h - b = \frac{4h}{L^2}\left(\frac{L}{2} - a\right)^2$$

$$b = h - h - \frac{4a^2h}{L^2} + \frac{4ah}{L}$$

$$b = \frac{4ah}{L^2}(L - a)$$

i.e. for a parabolic arch span $L$ height $h$, the height of any point P distance $a$ from one support is equal to

$$\frac{4ah}{L^2}(L - a)$$

Let a tangent to the parabola at any point P make angle $\alpha$ with the horizontal. This tangent will cut the $y$-axis such that OC = OD, but

$$OD = h - b$$

$$tan\ \alpha = \frac{2(h - b)}{L/2 - a}$$

$$= \frac{4h - 16ah/L^2(L - a)}{L - 2a}$$

$$= \frac{4h}{L^2}\left[\frac{L^2 - 4ah + 4a^2}{L - 2a}\right]$$

$$= \frac{4h}{L^2}\frac{(L - 2a)^2}{L - 2a}$$

$$= \frac{4h}{L^2}(L - 2a)$$

or the slope of a parabolic arch at any point P is equal to

$$4h/L^2(L - 2a)$$

## TYPES OF ARCH

There are three basic arch structures: the *three-hinged*, the *two-hinged* and the fixed arch as shown in Figs. 135(a), (b) and (c) respectively.

The forces in a three-hinged arch may be found by simple statical analysis. The two-hinged and fixed arches, however, are statically indeterminate and are dealt with in *Theory of Structures* in this series.

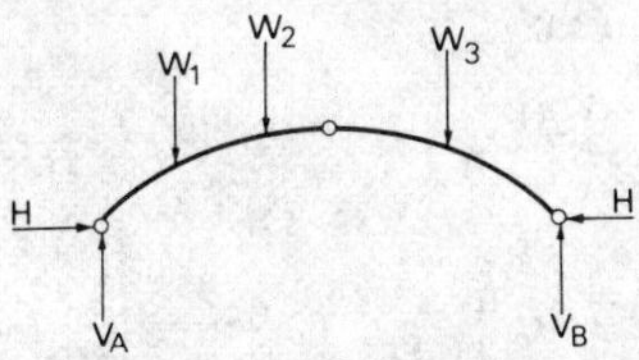

FIG. 135(a) 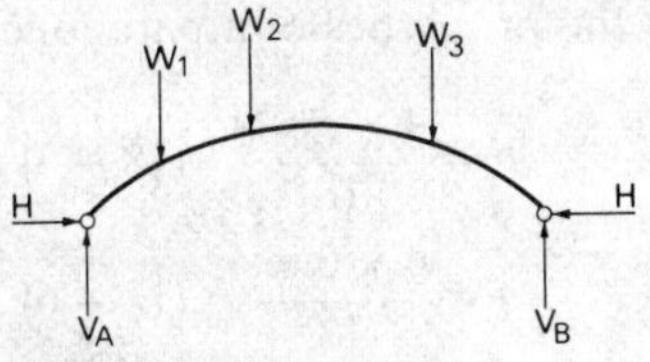          FIG. 135(b)

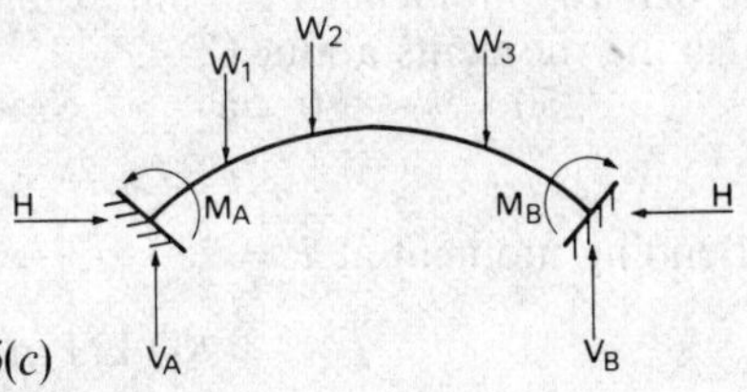

FIG. 135(c)

## 1. Three-hinged arch

A three-hinged arch is a statically determinate structure since an equation involving $H$ can be found by taking moments about the centre pin.

SPECIMEN QUESTION 63

A three-pin parabolic arch spans 18 m, and the hinge at the crown is 4 m above the springing. It carries point loads of 180 kN, 60 kN and 240 kN as shown in Fig. 136. What are the bending moments under the loads? Calculate the normal and shear thrust in the arch rib under the 60 kN load.

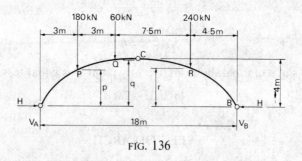

FIG. 136

SOLUTION

$$V_A = \frac{180 \times 15 + 60 \times 12 + 240 \times 4.5}{18} = 250 \text{ kN}$$

$$V_B = 230 \text{ kN}$$

From the properties of a parabolic arch:

$$p = \frac{4 \times 3 \times 4}{18^2} \ (18 - 3) \ = \frac{20}{9} \text{ m}$$

$$q = \frac{4 \times 6 \times 4}{18^2} \ (18 - 6) \ = \frac{32}{9} \text{ m}$$

$$r = \frac{4 \times 4.5 \times 4}{18^2}(18 - 4.5) = 3 \text{ m}$$

The bending moment at the pin at C must be zero.
Taking moments about C,

$$250 \times 9 - 180 \times 6 - 60 \times 3 - H \times 4 = 0$$

$$H = 247.5 \text{ kN}$$

Bending moment at $P = 250 \times 3 - 247.5 \times \frac{20}{9}$

$$= 200 \text{ kN m (sagging)}$$

Bending moment at $Q = 250 \times 6 - 180 \times 3 - 247.5 \times \dfrac{32}{9}$

$\qquad\qquad\qquad = 80$ kN m (sagging)

Bending moment at R $= 230 \times 4.5 - 247.5 \times 3$

$\qquad\qquad\qquad = 292.5$ kN m (sagging)

Under the 60 kN load

$\qquad$ Slope of arch, $\tan \alpha = \dfrac{4 \times 4}{18^2} (18 - 2 \times 6)$

$\qquad\qquad\qquad \tan \alpha = 0.296 \quad \therefore \quad \cos \alpha = 0.959$

$\qquad\qquad\qquad \sin \alpha = 0.284$

$\qquad$ Normal thrust in arch $= 247.5 \times 0.959 + 250 \times 0.284$

$\qquad\qquad\qquad\qquad - 180 \times 0.284$

$\qquad\qquad\qquad\qquad = 257.2$ kN

$\qquad$ Shear force in arch $= 250 \times 0.959 - 247.5 \times 0.284$

$\qquad\qquad\qquad\qquad - 180 \times 0.959$

$\qquad\qquad\qquad\qquad = -3.2$ kN

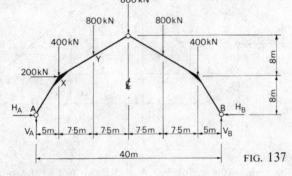

FIG. 137

SPECIMEN QUESTION 64

Dimensions of, and the loading on, a three-hinged arch are shown in Fig. 137. Calculate the horizontal and vertical components of the reactions at the hinged supports, and also the bending moments on the arch rib at the load points marked X and Y.

SOLUTION

In this problem due to the 200 kN horizontal load at X, $H_A$ will not equal $H_B$, but $H_A + 200 = H_B$. Otherwise the procedure is the same.

$V_A = \dfrac{400 \times 35 + 800 \times 27.5 + 800 \times 20 + 800 \times 12.5 + 400 \times 5 - 200 \times 8}{40}$

$\quad = 1560$ kN

$V_B = 3200 - 1560 = 1640$ kN

Moments about pin at C,

$$1560 \times 20 - 400 \times 15 - 800 \times 7.5 - 200 \times 8 - H_A \times 16 = 0$$
$$H_A = 1100 \text{ kN}$$

Bending moment at X $= 1560 \times 5 - 1100 \times 8$
$$= -1000 \text{ kN m} \quad \text{(hogging)}$$

Bending moment at Y $= 1560 \times 12.5 - 400 \times 7.5 - 200 \times 4$
$$- 1100 \times 12$$
$$= 2500 \text{ kN m} \quad \text{(sagging)}$$

(Note usual sign convention.)

SPECIMEN QUESTION 65

The dimensions and loading of a three-hinged arch structure are shown in Fig. 138. Calculate (a) the horizontal and vertical components of the reactions at A and B, and (b) the bending moment at the 600 kN load point.

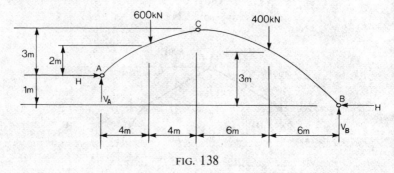

FIG. 138

SOLUTION

In this case the arch springings A and B are not at the same level. However for $\Sigma H = 0$ the horizontal reaction $H$ must be the same at A and B.

(a) Take moments about point B

$$20V_A - 600 \times 16 - 400 \times 6 + H \times 1 = 0$$
$$20V_A + H = 12\,000 \tag{1}$$

Take moments about point C
$$8V_A - 600 \times 4 - H \times 3 = 0$$
$$8V_A - 3H = 2400 \tag{2}$$

Solving equations (1) and (2)
$$\underline{\underline{H = 705.9 \text{ kN}}}$$
$$\underline{\underline{V_A = 564.7 \text{ kN}}}$$
$$V_B = 600 + 400 - 564.7 = \underline{\underline{435.3 \text{ kN}}}$$

(*b*) Under 600 kN load
Bending moment = $564.7 \times 4 - 705.9 \times 2 = \underline{\underline{847 \text{ kN m}}}$

## EXAMINATION QUESTIONS

1. Figure 139 shows the dimensions and loading of a three-pin parabolic arch rib. Draw a bending moment diagram for the arch marking on the values of bending moment under the load points. Calculate the normal thrust and shear force at the central pin and in the arch rib under the 200 kN load.

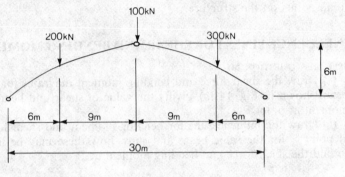

FIG. 139

2. Calculate the direction and magnitude of the resultant reactions at the pinned springings of the three-hinged arch shown in Fig. 140.

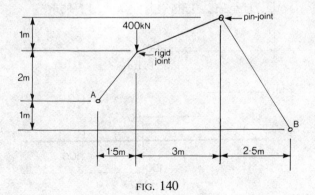

FIG. 140

# CHAPTER 7
# INFLUENCE LINES

An influence line shows how the value of a function (bending moment, shear, deflection, etc.) varies at *one given point on a structure* as a single unit load moves across the structure.

It is important to note that the load is moving and the influence line refers to a fixed point on the structure, whereas a bending moment or shear force diagram has a fixed loading system and refers to all points on the structure.

## INFLUENCE LINES FOR SHEAR AND BENDING MOMENT

SPECIMEN QUESTION 66

(*a*) Draw the shear force and bending moment diagrams for the beam shown in Fig. 141(*a*), giving the value of shear and bending moment at point P.

(*b*) Draw the influence line for bending moment and shear force at point P for the same beam and show how these may be used to find the shear force and bending moment at P.

SOLUTION

(*a*) The shear force diagram is shown in Fig. 141(*b*) and the bending moment diagram in Fig. 141(*c*).

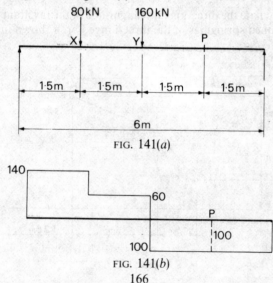

FIG. 141(*a*)

FIG. 141(*b*)

166

(b) For the influence line of shear force for point P consider a unit load, distance $x$ from the left-hand reaction as shown in Fig. 141(d).

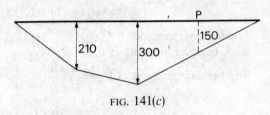

FIG. 141(c)

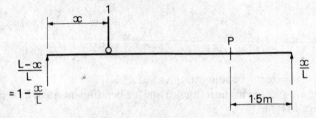

FIG. 141(d)

Shear force at $P = +\dfrac{x}{L}$ when unit load is to the left of P

$$= \dfrac{x}{L} - 1 \text{ when unit load is to the right of P}$$

both of these are linear variations. Also when

$x = 0$, shear force at P $= 0$

$x = 4\cdot5$, shear force at P $= \frac{3}{4}$ to the left

or $-\frac{1}{4}$ to the right

$x = 6$, shear force at P $= 0$

These are plotted in the influence line for shear force at P, shown in Fig. 141(e).

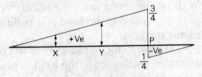

FIG. 141(e)

From Fig. 141(e), using similar triangles

ordinate at $X = +\frac{1}{4}$; ordinate at $Y = +\frac{1}{2}$

Therefore the shear force at $P$ with loading shown in question is

$$80 \times \tfrac{1}{4} + 160 \times \tfrac{1}{2} = 100 \text{ kN}.$$

For the influence line of bending moment at point P consider the same unit load.

Bending moment at $P = 1 \cdot 5 \dfrac{x}{L}$ when unit load is to the left of P

or $\qquad\qquad = 4 \cdot 5 \left( 1 - \dfrac{x}{L} \right)$ when unit load is to the right of

P, which are again linear variations.

When $\quad x = 0, \quad$ bending moment at $P = 0$

$$x = 4 \cdot 5, \quad \text{bending moment at } P = \frac{1 \cdot 5 \times 4 \cdot 5}{6}$$

$$= \frac{4 \cdot 5}{4} \text{ kN m}$$

or $\qquad\qquad\qquad = 4 \cdot 5 \left( 1 - \dfrac{4 \cdot 5}{6} \right)$

$$= \frac{4 \cdot 5}{4} \text{ kN m}$$

$x = 6, \quad$ bending moment at $P = 0$

These are plotted in the influence line for bending moment at P shown in Fig. 141($f$).

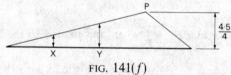

FIG. 141($f$)

From Fig. 141($f$),

ordinate at $X = \dfrac{1}{3} \times \dfrac{4 \cdot 5}{4} = \dfrac{3}{8}$; ordinate at $Y = \dfrac{2}{3} \times \dfrac{4 \cdot 5}{4} = \dfrac{3}{4}$

Therefore the bending moment at P with loading shown is

$$80 \times \frac{3}{8} + 160 \times \frac{3}{4} = 150 \text{ kN m}$$

For a simple beam with a fixed loading system it is obviously not advantageous to draw influence lines for shear force and bending moment at a point. If, however, the load is a rolling load, then this is a most convenient method.

To obtain the maximum value of shear, bending moment, etc. due to a single concentrated rolling load, the load is placed at the point where the ordinate to the influence line for that function is a maximum. If the rolling load is distributed the maximum value of the function may be found by inspection.

SPECIMEN QUESTION 67

A simply supported beam spans 8 m. Draw the influence line for shear close to one support and bending moment at the centre of the span. From these influence lines estimate the maximum values of shear force and bending moment in the beam with (*a*) a single load of 80 kN crossing the span; (*b*) a pair of loads of 50 kN and 30 kN fixed 2 m apart crossing the span; (*c*) a uniformly distributed load, 6 m long, of 10 kN/m crossing the span.

SOLUTION
In this question the critical points for maximum shear and bending moment for this particular structure have been chosen.

With unit load crossing the span distance $x$ from the left-hand reaction (Fig. 142(*a*)), the shear force influence line for the left-hand support would be as shown in Fig. 142(*b*) and the bending moment influence line for the centre of the beam as shown in Fig. 142(*c*).

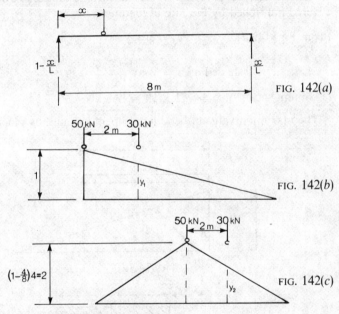

FIG. 142(*a*)

FIG. 142(*b*)

FIG. 142(*c*)

(*a*) With a single 80 kN load crossing span.

maximum value of shear force = $\underline{\underline{80\ kN}}$

maximum value of bending moment = $80 \times 2 = \underline{\underline{160\ kN}}$

(*b*) With a train of 2 loads 50 kN and 30 kN fixed 2 m apart.
For shear force, worst condition is when train is adjacent to support i.e. 1 load at support, other load 2 m into span

$$\text{ordinate } y_1 = \frac{6}{8} = 0.75$$

Maximum value of shear = $50 \times 1 + 30 \times 0.75 = \underline{\underline{72.5\ kN}}$

For bending moment, worst condition is when train is close to centre

$$\text{ordinate } y_2 = \frac{2(8/2 - 2)}{4} = 1$$

$$\text{Maximum value of bending moment} = 50 \times 2 + 30 \times 1$$
$$= \underline{130 \text{ kN m}}$$

(c) With a uniformly distributed load, 6 m long of 10 kN/m crossing the span.

The maximum value of shear force will be when the distributed load covers the maximum area of influence diagram. The value of maximum shear force for the point will then be the area of diagram covered multiplied by the rate of loading.

From Fig. 142(d)          $y_3 = \frac{2}{8} = \frac{1}{4}$

$$\text{shaded area} = 6 \times \frac{(1 + 1/4)}{2} = 3 \cdot 75$$

$$\text{maximum shear force} = 3 \cdot 75 \times 10 = \underline{37 \cdot 5 \text{ kN}}$$

The maximum value of bending moment is found by placing the

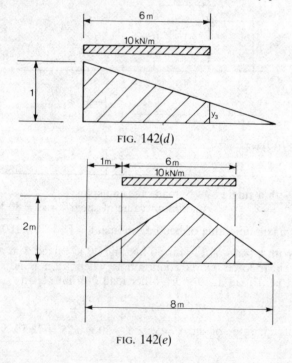

FIG. 142(d)

FIG. 142(e)

distributed load such that the area of influence line covered by the load is as large as possible. The value of maximum bending moment for the point will then be the area of the diagram multiplied by the rate of load. From Fig. 142(e)

$$y_4 = \tfrac{2}{4} = \tfrac{1}{2}$$

$$\text{shaded area} = 2 \times 3 \times \frac{(2 + 1/2)}{2} = 7\tfrac{1}{2}$$

maximum bending moment = $7\tfrac{1}{2} \times 10 = \underline{\underline{75 \text{ kN m}}}$

As may be seen it is not always a case of simple inspection to find the maximum value of bending moment, particularly if there is a train of concentrated point loads. However, from consideration of a simple bending moment diagram, it is apparent that the absolute maximum live moment must occur beneath one of the loads.

Influence lines may be plotted for a variety of functions on any type of structure, i.e. shear, bending moment or deflection of a beam, horizontal reaction of an arch rib or forces in lattice girders, etc. Some of these influence lines require a greater knowledge of theory than can be given in this volume and may be found in *Theory of Structures* in this series. A few simple cases are dealt with below.

## INFLUENCE LINE FOR HORIZONTAL THRUST ON A THREE-HINGED ARCH

A simple expression for the influence line for horizontal thrust in three-hinged arches is readily developed. Consider point P distance $nL$ from the left-hand support of a three-hinged arch, span $L$, height $h$, with unit load crossing the arch (*see* Fig. 143(a)).

With unit load at point P

$$V_{\text{B}} = \frac{nL}{L} = n$$

$$V_{\text{A}} = 1 - n$$

Taking moments about the central hinge from left-hand side

$$M = V_{\text{A}}\frac{L}{2} - 1\left(\frac{L}{2} - nL\right) - Hh = 0$$

$$\left(1 - n\right)\frac{L}{2} - \left(\frac{L}{2} - nL\right) - Hh = 0$$

$$H = \frac{nL}{2h}$$

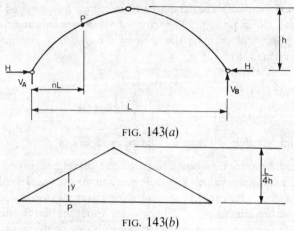

FIG. 143(a)

FIG. 143(b)

i.e. when $n = 0$, $H = 0$, when $n = \frac{1}{2}$, $H = \dfrac{L}{4h}$

from the right-hand side similar values are obtained.

The influence line diagram for horizontal thrust constructed from these results is shown in Fig. 143(b). For a number of loads crossing the span $H = \Sigma Wy$, where $y$ is the ordinate on the influence line diagram under the load.

At point P (Fig. 143(b))

$$\frac{y}{L/4h} = \frac{nL}{L/2}$$

or
$$y = \frac{nL}{2h}$$

Expressions for the influence line for bending moment, radial shear and normal thrust may also be developed.

SPECIMEN QUESTION 68

A parabolic three-hinged arch spanning 16 m, with a central rise to the hinge of 3 m, carries a system of two concentrated loads each of 50 kN and 1·5 m apart, which cross the span (*see* Fig. 144(a)). Calculate with the use of influence lines: (a) the horizontal thrust when the nearest load is 13 m from end A; (b) the maximum horizontal thrust for any load position; and (c) the resultant reactions at the supports when the nearest load is 5 m from end A.

SOLUTION

(a) For a single unit concentrated load,

$$H_{\max} = \frac{L}{4h} = \frac{16}{4 \times 3} = 1·33$$

The influence line for horizontal thrust is shown in Fig. 144(b).

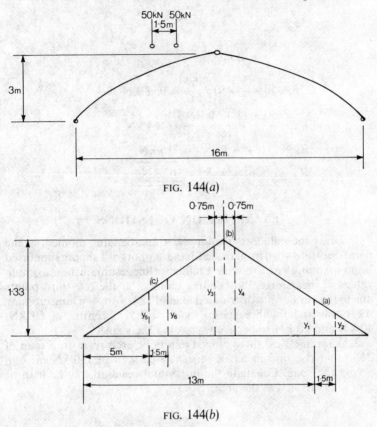

FIG. 144(a)

FIG. 144(b)

With nearest load 13 m from A

$$y_1 = \tfrac{3}{8} \times 1{\cdot}33 = \tfrac{1}{2}$$

$$y_2 = \frac{1{\cdot}5}{8} \times 1{\cdot}33 = \tfrac{1}{4}$$

$$H_A = 50 \times \tfrac{1}{2} + 50 \times \tfrac{1}{4} = \underline{\underline{37{\cdot}5 \text{ kN}}}$$

(b) The maximum horizontal thrust occurs when the loads are equidistant about the centre-line of the arch

$$y_3 = y_4 = \frac{7{\cdot}25}{8} \times 1{\cdot}33 = \frac{7{\cdot}25}{6}$$

$$H_{max} = \frac{2 \times 7{\cdot}25}{6} \times 50 = \underline{\underline{120{\cdot}8 \text{ kN}}}$$

(c) With the nearest load 5 m from A

$$y_5 = \frac{5}{8} \times 1 \cdot 33 = \frac{5}{6}$$

$$y_6 = \frac{6 \cdot 5}{8} \times 1 \cdot 33 = \frac{6 \cdot 5}{6}$$

$$H = 50 \times \frac{5}{6} + 50 \times \frac{6 \cdot 5}{6} = 95 \cdot 8 \text{ kN}$$

$$V_A = \frac{50 \times 11 + 50 \times 9 \cdot 5}{16} = 64 \text{ kN} \quad V_B = 36 \text{ kN}$$

$$R_A = \sqrt{(64^2 + 95 \cdot 8^2)} = \underline{\underline{115 \text{ kN}}}$$

$$R_B = \sqrt{(36^2 + 95 \cdot 8^2)} = \underline{\underline{102 \text{ kN}}}$$

## EXAMINATION QUESTIONS

1. Draw the influence line for shear and bending moment at the point one third span from the left-hand support of a simply supported beam spanning 9 m. From these influence lines estimate the maximum values of shear force and bending moment at the one third points due to (a) a uniformly distributed load of 12 kN/m, 4 m long crossing the beam and (b) three point loads at 1·5 m centres of 60 kN, 40 kN and 20 kN respectively, crossing the span.

2. A symmetrical three-hinged parabolic arch rib with a span of 24 m and rise of 3 m has a concentrated load of 180 kN moving across the span. Calculate the horizontal reaction for the load at 3 m intervals across the arch.

APPENDIX

# ANSWERS TO EXAMINATION QUESTIONS

## Chapter 1
1. 348 N/mm$^2$; 70·4 kN/mm$^2$.
2. Steel: 73 N/mm$^2$ tension, copper: 8·8 N/mm$^2$ compression.
3. 29·5 N/mm$^2$.
4. (a) 280 mm × 280 mm column with 4 No. 16 mm diameter bars.
   (b) 240 mm × 360 mm (rounded up to design sizes).
5. 0·00127.

## Chapter 2
1. 1·86 kN upwards to the right at 75° 22′ to the horizontal, passing 26 m vertically above D.
2. $AB$ = 36 kN tens.; $BC$ = 104 kN tens.; $CD$ = 143 kN compr.; $DE$ = 110 kN compr.; $DB$ = 92 kN compr.; $BE$ = 28 kN tens.
3. $AB$ = 80$\sqrt{2}$ kN compr.; $BC$ = 80$\sqrt{2}$ kN tens.; $CD$ = 20$\sqrt{2}$ kN compr.; $DE$ = 20$\sqrt{2}$ kN tens.; $EF$ = 100$\sqrt{2}$ kN tens.; $FG$ = 100$\sqrt{2}$ kN compr.; $BD$ = 160 kN compr.; $DF$ = 200 kN compr.; $AC$ = 80 kN tens.; $CE$ = 160 kN compr.; $DF$ = 200 kN compr.; $AC$ = 80 kN tens.; $CE$ = 180 kN tens.; $EG$ = 100 kN tens.
4. $AB$ = 0; $BC$ = 640/3 kN tens.; $CD$ = 640/3 kN tens.; $DE$ = 0; $FG$ = 160 kN compr.; $GH$ = 160 kN compr.; $HJ$ = 400/3 kN compr.; $JK$ = 400/3 kN compr.; $AF$ = 120 kN compr.; $BG$ = 0; $CH$ = 0; $DJ$ = 0; $EK$ = 100 kN compr.; $FB$ = 200 kN tens.; $BH$ = 200/3 kN compr.; $HD$ = 100 kN compr.; $DK$ = 500/3 kN tens.
5. (1) = 560 kN tens. (2) 1120 kN tens. (3) 613 kN tens. (4) = 240 kN tens. (5) = 280 kN compr. (6) = 840 kN compr. (7) = 866·7 kN compr. (8) = 360 kN compr. (9) = 120 kN compr. (10) = 350 kN tens. (11) = 350 kN compr. (12) = 350 kN tens. (13) = 350 kN compr. (14) = 316·7 kN compr. (15) = 316·7 kN tens. (16) = 316·7 kN compr. (17) = 150 kN tens. (18) = 150 kN compr. (19) = 150 kN compr.
6. $U_1A$ = 325 kN compr.; $U_1L_1$ = 180 kN tens.; $U_1L_2$ = 170 kN tens.; $U_1L_3$ = 156 kN compr.; $U_1U_2$ = 210 kN compr.
7. $R_A$ = 32·7 kN upwards, 2·7 kN to the left; $R_B$ = 37·3 kN upwards, 37·3 kN to the left; $F_x$ = 16 kN compr.; $F_z$ = 5·6 kN tens.
8. $AD$ = 200 kN compr.; $DE$ = 170 kN compr.; $EC$ = 105 kN compr.; $CF$ = 97·5 kN compr.; $FG$ = 115 kN compr.; $GB$ = 120 kN compr.; $AH$ = 173 kN tens.; $HJ$ = 121·5 kN tens.; $JK$ = 81 kN tens.; $KL$ = 95·4 kN tens.; $LB$ = 104 kN tens.; $DH$ = 52 kN

175

compr.; $EJ = 61$ kN compr.; $EH = 52$ kN tens.; $CJ = 53.3$ kN tens.; $CK = 19$ kN tens.; $FK = 21.7$ kN compr.; $FL = 8.7$ kN tens.; $GL = 8.7$ kN compr.

9. $1433 \times 10^6$ mm$^4$.
10. $598 \times 10^6$ mm$^4$.
11. B = 900 kg, C = 91 kg.
12. $AC = 43.1$ mm diameter.

## Chapter 3

1. $R_L = 95$ kN; $R_R = 85$ kN; $M_{max} = 420$ kN m under 100 kN load.
2. $M_{max} = 51$ kN m; 4.13 m from $A$.
3. $M_{max} = 125.8$ kN m; 2.93 m from left-hand support.
4. $M_{max} = 112.5$ kN m at D. Point of contraflecture 5.65 m from $A$.
5. $M_{max} = 125$ kN m, 2.9 m from left-hand support.
6. $R_A = 14.31$ kN; $R_B = 25.69$ kN; $M_{max} = -22.5$ kN m at $B$. In $AB$, $M_{max} = 11.5$ kN m, 1.73 m from $A$.
7. $R_B = 86.6$ kN; $R_E = 90.9$ kN; $M_B = -7.5$ kN m; $M_C = 83$ kN m (max); $M_D = 67$ kN m and 32 kN m. $M_E = -60$ kN m.
8. $M_A = -160$ kN m (max -ve); $R_A = 50$ kN; $R_C = 90$ kN; between B and C, $M_{max} = 67.5$ kN m.
9. $R_L = 20.4$ kN; $R_R = 290.6$ kN; Max span Mt = 108 kN m, 0.85 m from $A$. $M_{max} = -108$ kN/m at $B$; $f_x$ max = 158 N/mm$^2$.
10. $I_{NA} = 66.9 \times 10^6$; M.R. = 69 kN m.
11. $\omega = 14.5$ kN/m.
12. M.R. = 2.58 kN m (tension criterion).
13. As given $\omega = 11.75$ kN/m, flanges reversed $\omega = 23.4$ kN/m.
14. Max $f_t = 75$ N/mm$^2$; 6.54 m from $A$. Max $f_c = 25.8$ N/mm$^2$ at $B$.
15. Max $f_c = 130$ kN/m$^2$, Max $f_t = 90$ kN/m$^2$.
16. (a) Between 13.8 mm and 57.3 m from outside of flange.
    (b) Between 51.7 mm and $-32$ mm from outside of flange.
17. $\omega = 37$ Mg.
18. $P = 59.8$ Mg; $f_c = 7.15$ N/mm$^2$; $f_t = 0.64$ N/mm$^2$.
19. Max $f_c = 86.6$ N/mm$^2$; Min $f_c = 26.6$ N/mm$^2$.
20. (a) Max $p_c = 706$ kN/m$^2$; Min $p_c = 22$ kN/m$^2$.
    (b) Max $p_c = 698$ kN/m$^2$; Min $p_c = 34$ kN/m$^2$.
21. Max $f_c = 493$ kN/m$^2$; Min $f_c = 340$ kN/m$^2$. Maximum additional load = 3267 kN.
22. (a) $f_c = 18.8$ N/mm$^2$; $f_s = 31.5$ N/mm$^2$.
    (b) $f_c = 38.1$ N/mm$^2$ compr.; $f_s = 49.4$ N/mm$^2$ tension.
23. $f_c = 4.4$ N/mm$^2$; $f_s = 117$ N/mm$^2$.

## Chapter 4

1. $q_{max} = 24.5$ N/mm$^2$; $q_{mean} = 20$ N/mm$^2$ (based on web only).
2. $q_{max} = 62$ N/mm$^2$.
3. $q_{max} = 71$ F N/mm$^2$ (F in MN units).

4. $\omega = 68.2$ kN/m.
5. $E = 17.7$ kN/mm$^2$; $G = 7.3$ kN/mm$^2$; $\eta = 0.21$.
6. (a) $T_{max} = 1.33$ m kN.
   (b) $T_{max} = 2.45$ m kN.
7. min. dia. $= 223$ mm.
8. (a) $q = 37.6$ N/mm$^2$.
   (b) $p_2 = 21$ N/mm$^2$.
   (c) $q_{max} = 41.5$ N/mm$^2$.
9. (a) Max $f_t = 120$ N/mm$^2$ at 63° 26′ to the horizontal.
   (b) Max $f_c = 30$ N/mm$^2$ at 153° 26′ to the horizontal.
   (c) Max $q = 75$ N/mm$^2$ at 108° 26′ to the horizontal.
10. (a) $f_x = 68$ N/mm$^2$.
    (b) $f_x = 40$ N/mm$^2$.
    (c) $f_x = 33.5$ N/mm$^2$.
11. $f_{40} = 142$ N/mm$^2$; $q_{40} = 22$ N/mm$^2$.
    principal planes at 59° and 149° to plane of 46 N/mm$^2$ stress.
    $p_1 = 146$ N/mm$^2$ tens.; $p_2 = 10$ N/mm$^2$ tens.; $q_{max} = 68$ N/mm$^2$.

## Chapter 5

1. $\theta_A = -\dfrac{ML}{3EI}$; $\theta_B = \dfrac{ML}{6EI}$; max deflection $= \dfrac{ML^2}{9EI\sqrt{3}}$ at $\dfrac{L}{EI\sqrt{3}}$
   from B assuming moment applied at A.
2. $I = 171 \times 10^6$ mm$^4$; $d = 238$ mm.
3. $M_L = 278$ kN m; $M_R = 227$ kN m.
4. $R = \omega L/B$.
5. Moment at base $= 7.8$ kN m.
6. Moment at base $= 38.5$ kN m.
7. at centre, deflection $= 36.6$ mm; at 60 kN load, deflection $= 32$ mm.
8. $\delta_B = 49/EI$; $\delta_C = 64/EI$; $\theta_C = 29/EI$.
9. $\delta_B = 9.4$ mm.
10. At centre of AB, $\delta = -766/EI$; $\delta_C = 338/EI$; $\theta_A = 332/EI$; $\theta_B = 249/EI$.
11. $\delta = 16.4$ mm; 2.86 m from left-hand support.
12. $M_A = 243$ kN m; $M_B = 170$ kN m; $\delta = 232$ mm at 2.85 m from A.
13. $\delta = 116$ mm.
14. $M_A = 325$ kN m; $M_B = 425$ kN m; $\delta = 8.6$ mm;
    maximum $M = 316$ kN m at 5.3 m from A.

## Chapter 6

1. $M_{200} = 180$ kN m; $M_{300} = 540$ kN m; $F_c = 375$ kN; $S_c = 70$ kN;
   $F_{200} = 454$ kN; $S_{200} = 81$ kN.
2. $R_A = 305$ kN at 11° to vertical; $R_B = 97$ kN at 32° to vertical.

**Chapter 7**

1. (a) S = 21·9 kN; M = 64 kN m; (b) S = 70 kN; M = 203 kN m.
2. 0, 90, 180, 270, 360, 270, 180, 90, 0 kN.